Shimmer

Animalities

Books available
Robert Briggs, *The Animal-To-Come: Zoopolitics in Deconstruction*
Deborah Bird Rose, *Shimmer: Flying Fox Exuberance in Worlds of Peril*

Forthcoming books
Dominique Lestel, *Animality: An Essay on the Status of the Human*,
together with *Animalities* by Matthew Chrulew

Visit the series website at
edinburghuniversitypress.com/series-animalities

Shimmer

Flying-Fox Exuberance in Worlds of Peril

Deborah Bird Rose

EDINBURGH
University Press

Edinburgh University Press is one of the leading university presses in the UK. We publish academic books and journals in our selected subject areas across the humanities and social sciences, combining cutting-edge scholarship with high editorial and production values to produce academic works of lasting importance. For more information visit our website: edinburghuniversitypress.com

Edinburgh University Press Ltd
The Tun – Holyrood Road, 12(2f) Jackson's Entry, Edinburgh EH8 8PJ

Typeset in 10/14pt Warnock Pro and Gill Sans
by Cheshire Typesetting Ltd, Cuddington, and Cheshire

Series logo design by Anna-Katharina Laboissière

A CIP record for this book is available from the British Library

ISBN 978 1 4744 9038 2 (hardback)
ISBN 978 1 4744 9041 2 (webready PDF)
ISBN 978 1 4744 9039 9 (paperback)
ISBN 978 1 4744 9040 5 (epub)

Contents

Dedications

Peter Boyle, a poet gifted with generosity, has partnered with me in illness and wellness, pouring forth the supportive care that makes my life worth living. My gratitude is greater than my words, and I become conscious that nothing, absolutely nothing, exceeds poetry. Peter's multiple prize-winning work can be savoured silently, or it can be read aloud – both are effective. This book is dedicated to Peter from the depths of my love.

> Beside the narrow bed
> my night-light is staring right into me.
> I will hold your voice inside me as long as I can.
> When I sleep you'll go on walking
> through a steady explosion of white flowers.
> (From Peter Boyle, 'In the Small Hours', in *What the Painter
> Saw in our Faces*, Five Islands Press: Wollongong, 2001.)

* * *

Blunt Jackson is an extraordinary editor. We worked together on every chapter in the book, especially those towards the end, when it became something of a race as to who or what would finish first. Thanks to Blunt we were able to speed up the writing process, and some of the more eloquent parts of the book owe their existence to our shared work. Blunt is my beloved son, and in addition to being a great editor he follows a Tibetan Buddhist tradition known as Yuthok Nyingthig.

* * *

Artist, poet and therapist, my dear daughter Chantal Jackson has been with me throughout the illness and has been supportive in more ways than I can number. She brings a tenderness to all that she does, and the gentle hand in this book owes much of its existence to Chantal's lightness and humour.

Acknowledgements

This work was interrupted several times while I underwent difficult treatments for cancer. The research and writing spanned the years from 2009 to 2018. Numerous people gave of their time and insights, agreeing to be interviewed for the book or to read draft chapters and discuss ideas. My heartfelt thanks go to them all.

The flying-fox carers and advocates I interviewed were generous with their time, explaining and demonstrating their work, and reflecting on their experiences and on wider issues of values and commitments. They have waited a long time to see the results of their generosity, and the best I can say is that I am incredibly grateful to have met such fine people and to have learned so much from them while I was able to do so. I offer my particular thanks to Carol Booth, Nick Edards, Jenny Mclean, Tim Pearson, Naomi Roulston, Louise Saunders, Storm Stanford and Denise Wade.

The stories I tell in this book are also inspired and shaped by the Aboriginal families and teachers who have shared their lives and stories with me over so many decades. While most of these teachers are now dead, I am able to continue to thank them by keeping their stories moving in the world. Each time I go back to my notebooks, or to stories that I've told in one context that I want to return to in another, they just keep unfolding.

I am also indebted to the many academic communities that have nurtured my thinking and the colleagues who have thought with me. I am particularly grateful to Michelle Bastian, Jeffrey Bussolini, Matt Chrulew, Stuart Cooke, Rick De Vos, Natasha Fijn, Kathie Gibson, Donna Haraway, Jim Hatley, Emily O'Gorman, Val Plumwood, Isabelle Stengers and Thom van Dooren. The stimulating collaborations of the Extinction Studies Working Group and the Kangaloon group have guided me in this project from beginning to end.

A small group of gifted scientists working in allied areas have also inspired and challenged me; particular thanks to Arian Wallach, Dan Ramp, Marc Bekoff, and the other researchers working towards a Compassionate Conservation.

As always, this research would not have been possible without the love and support of my family. My children Chantal and Blunt Jackson, and my partner Peter Boyle. My brother Will, and my sisters Mary and Betsy. The legendary bushman, Darrell Lewis. Finally, my parents, Margaret and David Rose, both of whom were still alive when I began writing this book. My father's tender compassion for all vulnerable creatures has touched my heart all my life.

During the last four years of work on this book I was aided and comforted by the many wonderful staff at St George Hospital, in particular Dr Shir-Jing Ho. As I struggled through two rounds of cancer treatment and remission, they provided me with excellent care, frequently checking in on the progress of the book.

Portions of some of the chapters in this book have appeared elsewhere, and I acknowledge the earlier publication:

+ Rose, Deborah Bird. 'Flying Fox: Kin, Keystone, Kontaminant'. *Australian Humanities Review* 50 (2011): 119–36.
+ Rose, Deborah Bird. 'Cosmopolitics: The Kiss of Life'. *New Formations* 76 (2012): 101–13.
+ Rose, Deborah Bird. 'Multispecies Knots of Ethical Time'. *Environmental Philosophy* 9, no. 1 (2012): 127–40. DOI: 10.5840/envirophil2012918

• Rose, Deborah Bird. 'Shimmer: When All You Love is Being Trashed'. In *Arts of Living on a Damaged Planet: Ghosts and Monsters of the Anthropocene*, edited by Anna Lowenhaupt Tsing, Heather Anne Swanson, Elaine Gan and Nils Bubandt, G51–G63. Minneapolis: University of Minnesota Press, 2017.
• van Dooren, Thom and Deborah Bird Rose. 'Lively Ethography: Storying Animist Worlds'. *Environmental Humanities* 8 (2016): 1–17.

This research was undertaken with the financial support of the Australian Research Council (DP110102886) and with ethics approval from the Human Research Ethics Committee at Macquarie University.

* * *

Deborah Bird Rose passed away in December 2018. At the time of her death this book manuscript was almost entirely complete. Some minor editing and formatting were carried out by Thom van Dooren and Matthew Chrulew with the assistance of Sam Widin.

1   Speaking of Love and Peril

I sat in the backyard of a dilapidated bungalow in south-west Sydney wondering how I could possibly get used to living in this huge, dysfunctional city. I sat there often, brooding, and I continued to brood until the day the *Grevillea robusta* burst into flower. This large relatively nondescript tree, also known as silky oak, has a Cinderella moment when it decks itself out with bright orange, heavily nectared flowers. That evening I was sitting there as usual when suddenly I heard chattery quarrelsome sounds and smelled a glorious muskiness. I felt my senses come alive, and when I looked up I saw unmistakeable shapes in the darkening sky. I knew then that flying-foxes were abroad, bringing their blessings even into my backyard. I had known them when I lived far away in the tropics but had no idea they could become city dwellers. They offered me a glimpse of an urban 'nourishing terrain' in which multispecies gifts of life were on the move. 'Yes!' I thought. 'If you guys can live here, I can too!' A multitude of questions erupted.

They flew in for the nectar, but where was their camp? What exact species lived in Sydney? Were they nomadic, like the ones I'd known in North Australia? How and why had they become urbanites, and how were they faring? I soon learned that the flying-foxes which visited my backyard were 'Greys', and I was shocked to discover that they are listed as a Threatened Species (vulnerable to extinction). Greys are one of the four main species of Pteropids in Australia: Black Flying-Fox (Blacks, *Pteropus alecto)*, Grey-headed Flying-Fox (Greys, *P. poliocephalus*), Little Red Flying-Fox (Little Reds, *P. scapulatus*) and Spectacled Flying-Fox (Spekkies, *P. conspicillatus*).[1] During my years in the bush with Aboriginal people, I had come to know flying-foxes as Dreaming figures, as totemic kinfolk, as ecological communicators, and as food. They seemed abundant, and they were well loved and respected. My research in south-east Australia brought me into an altogether more conflicted topography. I learned that many flying-foxes were not faring well, and that many human beings held them in contempt and sought to eradicate them. In fact, the Greys which lifted the gloom from my urban life were the subject of special legislation that allowed them to be harassed, 'dispersed' and shot even though as native species and as species under the threat of extinction they would otherwise be protected.

I had just finished writing a book about love and extinction focusing on Australian dingoes. Now I was called to flying-foxes. My research questions led me into multispecies ethnographic work

Figure 1.1 Flying-fox in flight
Source: © Nick Edards.

involving wildlife carers and academically trained scientists in eastern Australia. The people I met were at the front line of the work of holding back flying-foxes from the edge of extinction. I continued to visit the north, and I revisited my notebooks from several decades of research with Aboriginal people. The research was exhilarating, and then again at times deeply disheartening. I was to encounter more passion, intimacy, cruelty, horror, complexity, generosity and wild beauty than I could ever have imagined. Living with flying-foxes, I came to understand, takes us straight to the heart of every big question facing Earth life in the twenty-first century, and in this encounter every bit of courage, empathy and compassion is called upon.

* * *

Australian flying-foxes are doing it tough. They are among the greatest flying mammals on Earth, and during their 40-million-year history in Australia they flourished thanks to deep and abiding mutualisms. With colonisation, European settlers decided that these creatures were pests and should be eradicated. So virulent was their animosity that in the early twentieth century the government brought a biologist from the UK to make the deathwork more effective. Flying-foxes survived decades of violence, alongside the more general destruction of their habitat, until more recent legislation granted modest protection as each of the four main Australian species is native. Now global warming has kicked in and flying-foxes are experiencing the greatest mammalian mass deaths in the world. Year after year. The story is bleak, but it is not only about death. Flying-foxes interact with humans in networks focused on kinship, care, rescue, advocacy and research. Human beings offer love and commitment; they bring assistance, bear witness and, through their actions, testify to an interspecies ethic that responds to suffering and, at the same time, praises the gifts that flying-foxes bring to the world.

These magnificent creatures are at the centre of this book, and from them my focus extends more widely to include many sustaining mutualistic connections involving humans, trees, waterways, and much more. What grabs me is the enthusiasm for life we encounter with flying-foxes. And further, that there are impassioned and compassionate responses of care and empathy within and across species borders.

Flying-foxes are nectivorous and seed-eating mammals and as such are key pollinators and seed dispersers. They bring gifts of life to Australia's patchy and increasingly fragmented woodlands and forests, giving and receiving, nurturing and being nurtured. As keystone species, their gifts keep flowing – through trees and into other plants and animals, into complex ecosystems, into oxygen and soil. Flying-foxes' partners in sustenance flourish because of the work flying-foxes do in the world. Humans who promote the well-being of flying-foxes become participants in these flows of gifts, and in this way both the short- and long-term life on Earth is enhanced.

Recent and contemporary human involvement with flying-fox life is far from uniform. Settler history in Australia is marked by massive violence. Ecocide and genocide were deeply entwined at the outset as settlers tried to remake the continent into their own vision, and while genocide is no longer an overt part of social life there continues to be majority acceptance of widespread war against certain species labelled as pests, and a relentless commitment to ecological destruction in the name of development.[2] With government approval, and even incentives, flying-foxes in the past were shot, poisoned, gassed, burnt and electrocuted. People have cut down the trees that housed their maternity camps, created a great variety of forms of harassment to drive them away, collected bounties for corpses and even bombed them. Of the four mainland species, two are in a spiral of decline that may end in extinction and are listed as threatened species. Their numbers are plummeting, and in spite of the fact that they are protected the east coast states of New South Wales and Queensland allow them to be legally shot.

Many Australians resist this deathwork, refusing to normalise or justify cruelty and mass death. Two main human groups –

Aboriginal families and networks of carers and advocates – defend flying-foxes as individuals and as species.

Aboriginal people lived alongside flying-foxes during their 65,000+ years of inhabitation of this continent. During these millennia people and flying-foxes shared foods and forests; their work, each species in its own manner, resulted in relationships of mutual benefit. In the sacred geography of Dreaming creation, flying-foxes are powerful world-shaping figures, and in everyday life they are totemic kin. This is not to say that relationships are invariably non-lethal; people eat flying-foxes, too, in the Indigenous way that combines love, respect and restraint with nourishment.[3]

Alongside Aboriginal kinship, a cohort of dedicated people is developing new practices of hands-on interactions with flying-foxes in need. Carers rescue individuals, foster orphans and make it possible for many flying-foxes to return to the bush and thus to continue to live in their own manner. These networks and alliances of people resist violence, assist the needy and take up public opposition to violence. Primarily they are volunteers – carers, advocates, academic scientists, community educators and veterinarians – who offer their time, energy and material resources in the service of rescue, conservation, public awareness and care. For every massacre and dispersal there is opposition. Cruelty often prevails, but not always. And yet care is offered with unlimited generosity.

Ethics and Earth life

Ethics precedes ontology, the philosopher Emmanuel Levinas tells us.[4] In lived experience this means that to be alive is to be enmeshed in the lives of others, and that to be enmeshed is to bear responsibility. The conditions that make life possible precede us; before we take our first breath we are already indebted; there is no life without ethics. It follows that the responsibilities pervading our lives also precede our writing and reading. Ethics precedes writing, we can say, and to bear witness to others is already to have been called, already to have become implicated in their lives and deaths.

For Levinas, ethics is the way of living 'in openness to the vulnerability of others'.[5] This openness entails a double command: 'Do not murder' is the first claim made upon me. Inevitably there is more: 'do not abandon' is the second claim.[6] The peril of others calls us to respond with care, to bear witness, to rescue and to keep faith even, and perhaps especially, in the midst of deathwork. If we cannot stop the killing, we can ensure that human violence will be known; it will be opposed, and its toxic justifications will be refused.

Levinas developed his philosophy in the shadow of the Nazi genocide, and he focused on human suffering. The dual command, neither to abandon nor to kill, affirmed the ethical responsibilities of bystanders as well as perpetrators. Today we live in the shadow of the Anthropocene, and the calls that command our ethical responses include those of nonhumans, and of the Earth itself.[7] Many species today experience direct persecution, and many more suffer the indirect impact of human actions, like habitat loss and climate change; many are slipping out of the world of life. In this era of mass extinction, losses accumulate over time and result in cascades of interconnected death. One inspired scholar proposed modes of ethics that are 'recursive, contingent, and interactive dramas of encounter and recognition'.[8] Today such dramas, arising from the call 'do not abandon', address us powerfully and terribly from the edge of the abyss. Along with care and other forms of direct action, there is the challenge of fidelity towards life: bearing witness expresses one's commitment to ensure that those under attack shall not be left stranded, abandoned, disregarded, and, further, that the people who offer care, love and respect will themselves be supported and honoured.

* * *

If we are to respond to the calls of others we must be able actually to experience those calls. There are many reasons why we do not hear others. Geographical distance may be a factor, but in the media-saturated social matrix of our lives we may also be overwhelmed by the numbers and diversity of calls. And there are certainly other

reasons too – we may not know what is happening, or our ignorance may be strategic. Perhaps these numerous factors lead to the
atrophy of our ethical senses, disrupting our ability to attend and
relate to the natural world.

In addition to individual responses, cultural factors are also
implicated and have deep roots for some of us (humans). The
immediacy of any particular animal-human issue rests on the deep
history of the West's nature-culture binary, which is also an animal-
human binary.[9] The logic is that of hyper-separation: that which
is on one side of the boundary can have nothing in common with
that on the other side. This is a difficult, indeed crazy, boundary to
try to enforce. We humans *are* animals, we *are* part of 'nature'. The
work of defending an impossible boundary is grounded in violence,
hatred and the infliction of extreme suffering. Dominique Lestel
writes of 'ontological hatred', arguing that a deep and abiding western antagonism towards animals is actually directed less towards
anything the animal does and more towards its very existence. One
strand of his argument concerns the curious mindset that wants
both to cause suffering to animals *and* to deny that animals actually
do suffer. It is no accident, Lestel contends, that the image of the
animal machine arose concurrently with the practice of vivisection.
And yet, the issues go deeper. He proposes that western societies
are constituted against nature in general and against the animal in
particular. It follows that in sustaining this position 'against' animals, some humans want to make animals suffer.[10]

The question of ontological difference is a philosophical quest for
what is 'proper' to humans. In philosophical usage this term aims
to identify the differences that make an irrevocably qualitative difference. Traditionally, various capabilities were defined as uniquely
human, and thus as proper only to humans: communication, consciousness, emotions, even the capacity to experience suffering were
at various times defined as boundary-making differences. Such differences were placed in a hierarchy of human supremacy, leaving
the animals with 'a cluster of inabilities'.[11] A key indicator has been
consciousness. Several related terms express this elusive capacity:

consciousness, sentience, cognition, mind and subjectivity are similar but not identical. When animals and other nonhumans were categorised as mindless matter, such questions did not arise. The term cognition is a good example. It refers to a creature's ability to receive sensory input, transform it into knowledge, and apply that knowledge.[12] In short: input – receiving sensory evidence; knowledge – making something of that evidence; action – putting knowledge to work in the world. A vast amount of new scientific research shows that animals and plants live through cognition. Indeed, Lynn Margulis and Dorion Sagan hold that mind (synonymous with cognition) is pervasive within all life.[13] While immensely widespread, cognition takes different forms in different species. My focus on flying-foxes brings us into a kind of cognition best defined as subjectivity. This means that individuals are aware of themselves as thinking and acting subjects (not simply the objects of the thought of other species); furthermore, as members of their own species or kind, flying-fox subjectivity involves culture. Individuals inhabit and shape their life worlds (Chapter 4), do their best to ensure intergenerational continuity, and are vulnerable not only to physical suffering but to mental and emotional suffering as well (Chapters 7 and 8).

Debates about a singular 'proper' of humanity will continue, but some scholars are making the more interesting move of simply turning towards diversity. Recent work in science and technology studies,[14] multispecies ethnographies,[15] biosemiotics,[16] philosophy,[17] ethology[18] and etho-ethnology, to take a few examples, starts from entirely different premises. Dominique Lestel and his colleagues offer a particularly fine exposition of the current state of research in the interdisciplinary field of etho-ethnology. Their approach is to work with multispecies communities in which members are subjects, and relationships are intersubjective, that is between subject and subject. They note that difference can be reimagined as behavioural style:

One chimpanzee can be so different from another that its description reduced to identification of the species sheds very little light on its behaviour . . . Nevertheless it is still true, and we must be well aware of the fact, that there will always be a cognitive or behavioural style that will characterize chimpanzees as chimpanzees and distinguish them as much from gorillas as from elephants and humans.[19]

From this perspective, there are many 'propers'. A species will have its own distinguishing features, and there may also be smaller subgroups that share their own distinguishing propers: song communities, clans and numerous other creaturely genres. In Lestel's analysis of the diversity of propers there is no necessary hierarchical ordering and no push towards supremacy. Rather, he proposes, the structure of diversity is heterarchical, meaning that interactions flow across horizontal planes rather than moving up or down hierarchical ladders. We will see this structure of difference in, for example, the mutualism between trees and flying-foxes.

* * *

The remainder of this chapter introduces some of the key approaches, understandings and commitments that guide this book.

Ethos: In the context of care, I will be using the term 'world' while focusing primarily on individual flying-foxes. The idea that nothing comes without its world defines the term world as a lifeway drawn from the conjunction of body, self and environment, along with the subjectivity that holds it all together. When creatures share their type of body, mode of selfhood, environments and cultures we can discern patterns of a biocultural matrix. Such a matrix can be understood as an 'ethos' (plural ēthea).[20] The term comes from old Greek, where it meant things like character or way of life, but also custom, and customary practices and places. Although not widely used today, the term retains a place in anthropology where the focus is on humans: 'A people's ethos is the tone, character, and quality of their life, its moral and aesthetic style and mood; it is the underlying attitude toward themselves and their world that life reflects.'[21] An ethos is what makes a group or 'kind' distinct, and this distinctiveness takes many different, but interwoven, bio-cultural forms.[22] Behavioural and cognitive forms of distinctiveness are a central part of what constitute the ēthea of particular animal beings. An ethos is an *embodied* way of life; a way of reproducing, of forming social groups. It is everything that together constitutes a distinctive 'way of being'. Ēthea are shared within a group, whether that be a small local group, a species, or a related set of species. There is, for example, an ethos that shapes the lives of members of each species, and there is an ethos that is shared among the four main species of Australian flying-foxes. The goal of ethos analysis is not to solidify differences, or to propose a straightforward correlation of ēthea with species, but rather to focus our attention so as to recognise and respond to these differences, and thus to appreciate the fact that many nonhuman others really do live within biocultural worlds that have histories and that are shaping worlds and lives.

The fact is that we will never know what it is like to be flying-fox, or a tree. As the ecologist Frank Egler famously, and wisely, put it: 'ecosystems may not only be more complex than we think, they may be more complex than we *can* think'.[23] What is true of ecosystems is true at other scales. Flying-foxes individually are more complex than we think, and complex in ways beyond our thought. Their ethos includes their many social skills and cultural repertoires. And yet we also share glimpses of worlds, actions and connectivities. The mutualisms that sustain all of us are not obscure, and new information is always emerging.

Responsibilities: One of the most devastating effects of the animal-human binary has been the rejection of the idea that we have ethical responsibilities towards other creatures. Although in recent years this binary increasingly has been undermined in favour of connectivities across borders of difference, there is still a strong social/political 'common sense' position that puts human interests above all others.

And yet, the call into responsibility is not dependent on the specifics of any given creature, its species, its usefulness, its cuteness; rather it is enough to know that the call is there. But at the same time, responses must always be appropriate to the needs of others, as best we can understand them.[24] In writing, thinking and working across the boundaries of species we find ourselves face to face with both mystery and familiarity. Others are not replicas of us, and at times the gap is incommensurable. And yet, we are all kin within the family of life on Earth. This insight into kinship was the 'real scandal' of Darwin's work; it reveals the connectivities that the animal-human binary sought to conceal.[25] When we live ethically, we become participants in flows of mutual life-giving. Ethics arising in the actual conditions of life cannot be abstract and universal, nor can they constitute a closed system.[26] By the same logic, ethical writing requires openness both to the peril and to the joy of others. There are words of alarm – necessary and passionate, aiming to amplify ethical claims. Equally there is praise and celebration – for the lives of others, their passion and their gifts. Even as I raise my voice against violence, I focus my study on the beauty of flying-foxes' ways of living: their high-flying verve, their joyful labour awash in pollen and nectar, their travels and attachments to home place, and their intensely social lives.

Multispecies ethnography: New understandings of connectivity enable new fields of research and writing that embrace affirmations of participation. One of the great anthropologists of the Anthropocene, Anna Tsing, evokes the excitement of this gripping moment:

> There is a new science studies afoot . . . and its key characteristic is multispecies love. Unlike earlier forms of science studies . . . it allows something new: passionate immersion in the lives of the nonhumans being studied. Once such immersion was allowed only to natural scientists, and mainly on the condition that the love didn't show. The critical intervention is that it allows

learnedness in natural science *and* all the tools of the arts to convey passionate connection.[27]

Deeply attentive to the lives of nonhumans, this new research is committed to engaging with diversity amongst humans as well. Multispecies ethnography, articulated initially by Eben Kirksey and Stefan Helmreich, is only possible because so many boundaries are now understood to be porous.[28] Wide-ranging, open and inclusive modes of research cultivate arts of attentiveness. Multispecies research brings us into encounter with 'a lively world in which being is always becoming, [and] becoming is always becoming-with'.[29]

You are not alone: The West's former view that all that was not human was simply mindless matter seems barely credible anymore.[30] And yet, a huge shift is required when we consider that our human lives are situated in vast realms of sentience. The Australian Aboriginal philosopher Mary Graham states that one of the basic premises of the Aboriginal worldview is: 'You are not alone in the world.'[31] And herein lies a powerful, perhaps alarming, challenge. In the midst of all this sentience, there is no hiding. The consequences of human action are not borne by mindless machines but by living beings, many of whom are conscious of their own lives and of the lives of others. And so, given that almost all the factors driving two Australian flying-foxes to the edge of extinction are biocultural (and include human and nonhuman actions), we bear responsibilities that are witnessed not only by other humans but by other living (and perhaps non-living) beings as well. We are called, therefore, into participation and intra-action.[32] It is true that, for better or worse, we always participate in life's flow (Chapter 7).

Connectivities: The science of ecology is wonderfully instructive. One of its central insights concerns ecological connectivity, defined as flows of energy and information across boundaries of difference. Difference attracts flows, and porous boundaries make flows possible. As connectivity is essential to life, so life necessarily entails flows, and therefore entails *inter*dependence. To put it in more spicy terms, connectivity reveals our position as participants in entangled co-becomings: nothing stands alone, everything, at pretty well every scale, depends on others through flows of energy and information.[33] Flows bring us directly into thinking with mutualisms – giving and receiving – and into exchanges that work across human and non-human domains and kingdoms. Many of the flows of energy and information bring forth and sustain life. Necessarily as well as beautifully, they are always in motion.

Creaturely bodies, different and same: There is ongoing discussion of how best to refer to our fellow beings, whether plant, animal or other. In general, I use the term creature, and I mean it to encompass all of us earthlings. The term derives from the same root as creation, and it thus connects us all, without favour towards any particular order or kingdom, as members of life's creative work in bringing forth diversity and complexity within the great family of life. Most of us probably associate the term creature only with animals, and for the most part, with my focus on the flying-foxes, that is how I use it. But when we consider the interactions between the animals and the trees they pollinate, it is actually more helpful to think of this as an inter-creature relationship rather than strictly as an animal-plant relationship (Chapter 2).

I learned the phrase 'different and same' from Aboriginal people, and it struck me forcibly because it is such a wonderfully succinct expression of the logic of both-and. This logic contrasts with the standard western logic based on either-or. The either-or logic posits difference as a problem to be overcome. Both-and logic focuses on relationships; it is inclusive and does not pose difference or sameness as problems to be resolved one way or another. The differences between these two logics may seem trivial at the outset, but both-and logic underpins a great deal of the ontological-ecological terrain of flows, waves and diversities; of borders that are porous rather than impenetrable. Dreaming creation, for example, and creation more generally, is both then and now; Dreamings in their ongoing creativity are both here and not here. Both different *and* same: we will see again and again that the inclusive complexity of this logic brings us into living worlds of *both* change *and* stability (Chapters 4 and 5).

Flame of life: Scientific approaches which emphasise the inter-connection of geophysical and biological processes, from Gaia theory to the more tame and seemingly controllable Earth Systems Science, describe a self-creating biosphere; life produces conditions that favour its own flourishing. We know this most impressively from the geological record. Across 3.5 billion years, every period of catastrophic loss is followed by exuberant bursts of coming forth. One of the most interesting scholars to address biosphere science within more poetic language is Eileen Crist. She uses the elegant term 'flame of life' to describe the recurrent process of life's recovery from catastrophe. In her words, the 'flame of life' arises from biosphere desire in modes of diversity, complexity and abundance.

These three qualities – diversity, complexity and abundance – show us the manner, or 'way', in which Earth recuperates itself after disaster. We can see and appreciate the long-term processes, and so we know from the perspective of western science how we humans can actively fit into working with earth systems and contribute to the robustness of the biosphere that sustains us. Other traditions of knowledge arrive at understandings of life's participatory coming forth in different languages and concepts (Chapter 4, for example), but the main point holds good: life desires more life. The three qualities Crist identifies as integral to the flame of life 'form the matrix of Earth's life-generating creativity and of the biosphere's robustness'.[34]

And yet, the evidence all around us in this Anthropocene era of mass death and destruction (Chapter 7) tells us that collectively we humans have become incredibly destructive. Crist puts it this way: 'human beings have taken aim at the very qualities that define the living planet, dismantling, with an intent that seems paradoxically both blind and demonic, the diversity, complexity, and abundance of life on Earth'.[35]

Ontological-ecological terrains: Flying-foxes live their lives within their own multispecies life worlds, and the stories of how they live and how they die offer one way into an exploration of an ontological-ecological terrain.[36] Other creatures and life worlds would be equally fascinating, I suspect, but having been called to flying-foxes (Chapter 4), I follow them in their glory and their peril.

With the term 'ecological' I am looking towards connectivity and flow. Thinking ecologically means learning to think with, and from within, the world. It contrasts with thought that works from the outside, over and above and against the world. With the term 'ontological' I am asking about the qualities of reality – the actual realness of the real, the foundational real that is the bedrock of the biosphere. The term 'terrains' expresses the fact that both ecological processes and ontological bedrock are situated in place and time. We are not talking about abstractions, here, but rather about reality in its fundamental manifestations through organic and inorganic domains, across processes such as intergenerational waves of life that are situated and interactive. In these waves of life coming forth into life worlds, and contributing to future waves (Chapter 6), there are always interactions between specific creatures that share mutual interests across boundaries of difference. There are circuits of flow that are local and intimate, and that are connected into wider and wider circuits that ultimately form the substance and continuities of life on Earth. 'Terrains' thus invites us to remain emplaced and avoid abstractions. At the same time, terrains may be nested. Connectivities exist at numerous scales: the planet, the regions, the actual places of our lives, and the connected places that ensure that no place is isolated or without significance. In scientific language, these circular, synergistic, life-giving relationships call for a concept of 'active fitting'.[37] Quite literally these terrains are the ground of being in a moment-to-moment sense, and they are also the ground of all the coming forth. Responsibility, love, care, commitment: they spring up from these grounds; they captivate us, they are the meaning of life.

To be sure, connections are not necessarily benign.[38] In relation to flying-foxes, there are terrible examples of human rapacity, hatred, domination, violence and cruelty (Chapters 7 and 8). Such is the social reality of our era, but it is not the only story, and it most certainly is *not* our ontological condition.

What do philosophers of ethics and goodness say about an ontological bedrock in which ethics actually precedes ontology? James Hatley wrote about Levinas's great dictum, and he invites us to think of creation (by which he means beginnings or origins) in two modes: both the originary moment of coming forth, and the ongoing processes of life.[39] Between these two large scales is the region within which we actually live, where creation is the ongoing process of coming forth in time and place. Unlike a mechanistic vision of creation in which the foundation is laid down at the beginning and the whole thing ticks along in closed and repetitive circuits, Hatley reminds us that in our everyday lives, our participation in flows and our precarious balance between birth and death, we Earth creatures are brought into life as inheritors of and participants in multispecies relationships of nurturance that vastly precede us (Chapter 6). The ontological-ecological terrains of our lives are at best embedded in generosity, responsibility, beauty and goodness that we ourselves did not make and towards which we have our own responsibilities. We are ethically engaged in the ongoing-ness of life before we even learn to understand and work with these deep commitments. We are called both by specific lives and by biosphere processes, and that call is for us to witness and to act, to become participants in the reality of life.

Hatley tells us that this actualisation of ethical life is Earth's ontological bedrock, and in our active participation we engage with the very realness of the real. This ontological insight concerning the primacy of ethics leads me to envision a human conscience that is shaped by its responsiveness towards the calls of others – nonhuman as well as human – and by connectivities that loop through organic and inorganic domains. This is a human conscience situated within the goodness of creation, and able to withstand the horrors of contemporary violence. It affirms its opposition to a world of wilful and deathful bloodshed. It is a difficult terrain; indeed, as Levinas said, we are called to exercise a difficult freedom.[40] And our position of witness requires that although we don't know what may yet happen, we still must act.

Yes!: Here is the great powerhouse of life on Earth. With 'Yes!' life moves, it acts, it comes bursting forth. Amongst flying-foxes, as with many humans, I encountered an embrace of the goodness of life's coming forth that constitutes a resounding 'Yes!' This affirmation conveys the desire to participate in the great flow of life's own desires and ways of becoming. 'Yes!' travels, it is interactive and relational, and it invites us to investigate how affirmations circulate. We know that 'Yes!' is affirmative: life comes bursting forth into relationships, and life wants to flourish. 'Yes!', with all its implied enthusiasm, guides us to life's pervasive, recurring 'tendencies of becoming': its diversity, complexity and abundance. For humans, saying yes to life is a profound ethical choice. It is a passionate embrace of the living world, a grateful response to the gifts of life, a pledge of solidarity with Earth life, and a passion to take up our share, each of us, in the complexities of mutuality. 'Yes!' is our continuing commitment to ontological-ecological terrains of reality: the affirmation that conveys our desire to participate in the great flow of life's own desires and ways of becoming.

Knowledge is always situated: We learn this in social terms from feminist thinkers who have analysed the West's epistemological practices. In Donna Haraway's incisive words, the idea of a disembodied, un-placed, all-knowing vision of knowledge is a 'God-trick' and is utterly false.[41] We also learn about situated and partial knowledge from quantum theory, and other recent discoveries in scientific knowledge. Physicists, for example, tell us that uncertainty is a fundamental condition of life.[42] Indeed, the more we learn about widespread sentience, the more we realise that knowledge is a startlingly uneven terrain. Every living thing is a parcel of perceptions, and every such parcel is situated amongst others in the folds and pleats not only of space-time but of information networks. There are areas of open flows of information and there are areas where little or nothing is known. This patchy terrain ensures that knowledge is always partial and provisional, and of course Earth is full of tricksters, as well. We act on hunches and guesswork, along with plans and hopes, and often we act in ignorance. Such is the uncertainty, perplexity and astonishment of life.

Because knowledge is situated, so it arises in interactions between creatures and contexts. This book is no exception. There are patchy bits, edges and irregularities resulting in a bumpy terrain that occasionally refracts my experience of the cancer that overtook my writing. My excellent doctors knew my determination to complete the book and did all they could to keep me going, checking from time to time on how the writing was progressing. And so, this book maps my experience of writing at the edge of my own limited span. Precisely because I face the unknown, the book explores in-between places of interactions marked by love, compassion, generosity and care. With flying-foxes and their mutualists as focal lenses, I encounter widespread stories of desire, trust and affirmation. I have written from and towards my love of the beauty of life, and in the end I hand the work over to readers in the hope that they will continue to expand the ripples of connection.

2   Meet the Pteropids

Australian flying-foxes are members of the *Pteropus* genus, one of several groups of predominantly fruit and blossom eating 'megabats' found in many parts of the world. These megabats, alongside their 'micro' cousins, make up the larger order Chiroptera, a group of animals known to most of us simply as bats. Chiroptera means hand winged, and the Chiropterans (mega and micro) are the only flying mammals in existence. The wing, a mammalian forelimb adapted for flight, bears the structural traces of its life as a hand or a paw.[1] The Chiropterans are diverse, extensive, numerous and of ancient lineage. One or more species inhabits every continent on Earth except Antarctica. One in every five species of mammals is a Chiropteran, and they are the most abundant of mammals.[2]

These unique creatures entered the evolutionary story in the Eocene, about 52 to 50 million years ago. Chiropterans have slow life cycles, are relatively long lived, and produce small numbers of offspring.[3] They almost certainly evolved from gliding mammals.

There are some outstanding differences between the megabats and the microbats. Megabats do not echolocate, but rather navigate by sight. It is likely that flying evolved first amongst the bats, with echolocation developing later in some smaller species.[4] Megabats are phytophagous: they eat plants, pollen, nectar, leaves and fruits.

Microbats, by contrast, include carnivorous, omnivorous and hematophagous species. Many of the micros live on insects, but some eat fish and other vertebrates; popular culture seems particularly fascinated by the hematophagous, or 'blood-eating', species.[5] The size of megabats, although variable, makes them easy to distinguish. The largest megabat, and therefore the largest flying mammal, is the 'greater flying-fox', *Pteropus giganteus*, ranging across India, Pakistan, Burma and beyond. Its wingspan is up to 170 cm (5'6").[6]

One theory has it that flying-foxes are actually descended from early primates, which would make them distant cousins of ours. According to this theory, little lemur-like creatures' front legs developed into wings and Pteropids took to the sky.[7] This theory has not withstood the test of DNA investigation very well, but nor has it been completely dismissed. When I first heard about the theory I felt a surge of recognition. A close encounter with a flying-fox induces the strong sense of being in the company of an odd little kinsman. With their small furry bodies and dog-/human-like faces, with their chattery camps full of individuals who are grooming each other and carrying on their daily life – mating, raising babies, guarding teenagers, bickering, competing and remaining attentive to sources of food in the region, it is difficult to understand how anyone could fail to recognise marvellously engaging kinfolk. And perhaps that is part of the problem in some places and at some times for some humans: flying-foxes are so close, and yet they are so different.

* * *

The mega branch of the Chiropterans probably originated in the region of Southeast Asia and Melanesia.[8] It includes forty-two genera with 166 species, including blossom bats, tube-nosed bats, dog-faced or short-nosed tube bats, pygmy fruit bats, musky fruit bats, and many more. This far-flung assemblage is spread out across Africa, the Middle East, and islands and costal continental regions across the Asia-Pacific. The subgroup that concerns me, the Pteropids, share this origin but did not move into Africa or the

Middle East; rather they kept to islands and the very edges of continental areas from Samoa to Zanzibar.

Dusk slips across the Old World tropics

The earth turns, day follows night, night follows day, the procession is so familiar and unvarying that we take it for granted. But in between these big everyday events there is a mysterious moment, a time of shifting transitions that is neither day nor night. Dusk and dawn are liminal – their swift passage speaks of shadows, offering fleeting glimpses of life in edgy places, and intimating the presence of things that belong not quite to the day or the night. With dusk, in particular, there is an element of mystery: the crepuscular, or twilight, moment is inhabited by creatures who revel in the half-light. Here, amongst dim, shadowy and yet expressive action, interactions resist easy definition. Dusk tells us that there will always be more to Earth life than either day or night can contain.

Dusk is liminal and flying-foxes are also liminal in their own way: they are so clearly mammalian, so clearly at home in flight, and so comfortable hanging upside down in their lively sociable camps, so similar to other animals, including humans, and yet so different. Across island and coastal regions of the Pacific and Indian oceans, dusk slips along, and with it comes the great Pteropid flyout. I imagine a wave in the specific sense of the term: a disturbance that moves across a physical reality that remains in place. At each home site there is noisy conversation accompanied by lots of pushing and shoving. In the half-light, one listens acutely: the rustling becomes more intense; the sound of wings and chit-chat becomes more active, and after a few tentative starts the mass rises up and spreads out.

The twin event – dusk and flyout – starts at the eastern edge of the Pteropid homelands in Samoa, and moves on to islands such as Fiji, Guam and the Marianas; it moves on to Japan, Taiwan and the Philippines. Parts of coastal China, and islands in the South China Sea, the Arafura Sea and the Timor Sea: each place experiences its own flyout. Australia, Papua New Guinea and Indonesia are brushed with dusk and flying-foxes. Across Southeast Asia – Malaysia, Cambodia, Vietnam, Thailand, Burma – and into parts of Nepal and Pakistan, India and Sri Lanka: still dusk travels, and still the flying-foxes respond. And so it goes, across the vast Indian Ocean to the Seychelles islands, Mauritius and Madagascar. Dusk moves across the whole world, but flying-foxes are limited in their travels. Beyond the islands near the coast of Tanzania there are no more Pteropids.

Given that flying-foxes are arboreal and nocturnal, their lives and human lives are potentially non-confrontational. However, flying-foxes eat fruit, perhaps not by preference but certainly where available; humans who grow fruit don't want to lose their crops. Conflict can become so severe that in many places persecutions, even massacres, rather than co-existence, have become the norm. Of the sixty or so species of Pteropids that recently lived in this vast region, several are now extinct, some are functionally extinct, and many more are threatened. Throughout this region of the Old World tropics conflict between humans and flying-foxes is on the rise.

The Australian contingent

Flying-foxes need water, and Australia is a challenge for them. Water in Australia is governed ecologically by the fact that this continent is the 'driest, flattest, most poorly drained, and in fact largely inward draining land on Earth'.[9] About 75 per cent of the continent is 'acutely arid' (of which 40 per cent is desert), 10 per cent is seasonally arid and acutely arid in some years, while only 15 per cent is 'reasonably well-watered'. Water in the desert is extremely unpredictable on an annual basis, and is influenced across the continent by the El Niño-Southern Oscillation (ENSO). Locally known as El Niño, this huge trans-oceanic climate dynamic amplifies unpredictability, brings high rates of variation, and is not linked with an annual cycle. It is cyclical, but the time frame is more likely to be a decade than just one year. El Niño gives rise to patterns of extreme oscillation between drought and flood that some ecologists label a 'boom and bust ecology'.[10] In Mary White's delightful term, it is a 'climate-warper'.[11] One result is that Australia is the driest and most unpredictable inhabited continent on Earth.

Most flying-foxes live in the relatively well-watered coastal and near inland areas; they can be encountered in Victoria, New South Wales, the Australian Capital Territory, Queensland, The Northern Territory and the northern part of Western Australia. Some have ventured into South Australia, and a few are even turning up in Tasmania.[12]

In Australia, the largest male flying-foxes weigh about one kilogram and have wingspans of up to 1.5 metres (5 feet).[13] Blacks are the largest, and until recently they lived primarily in the tropical and semi-tropical zones, with full-time inhabitation of some coastal areas, and seasonal trips inland to visit the flowering trees of the great savannah woodlands of the semi-arid regions. With global warming Blacks are extending their range further south where the winters are not as cold as they used to be. Greys live primarily along the south-east coastal region. Their range coincides with the most densely settled region of Australia, a region in the throes of immense development, so Greys are hardest hit by clashes with humans. Spekkies, so called because the rings around their eyes look remarkably like spectacles, are rainforest creatures; their Australian range is limited to the tropical east coast of north Queensland. Little Reds, Blacks and Spekkies also range beyond Australia into Papua New Guinea and Indonesia.[14] Little Reds are noticeably smaller than the others, as the colloquial name implies, and differ in other aspects as well. Their breeding cycle is six months out of sync with the other three, so while Greys, Blacks and Spekkies give birth in October–November, Little Reds give birth in April–May. Curiously, they have a noticeably different odour. Experienced people know just by the smell if Little Reds are in a camp, and Aboriginal stories, too, address the different smell. Little Reds camp together in even more closely crowded groups than the others, and they are not so constrained in their range, travelling further inland than the others, and venturing out into the deserts during periods of plenty. According to one academic scientist: 'The Little Reds go out there, but they're only following flowering, and they're pretty perspicacious about when they go where.'[15]

All the Australian Pteropids are arboreal, nocturnal, nomadic and live primarily on nectar, blossoms and fruit. They love to camp together in large numbers, but they do not all travel together all the time. Rather, as individuals they cover vast areas during an annual round as they follow flowering and fruiting trees and shrubs, and they return to home camps near large concentrations of food at specific times of year. They have a strong sense of site fidelity (philopatry), returning to maternity camps, in particular, year after year (Chapters 5 and 9). There are six main requirements for flying-fox life: water, food (including various elements such as salt), moderate temperature, tall trees, socialisation with each other (notably in home places), and the birth and care of new generations.

Flying-foxes are being monitored at a national scale by the Federal Department of Environment and Energy. There needs to be a more uniform approach to management of human/flying-fox conflicts, according to the scientists who work at the front lines.[16] A number of excellent academically trained scientists carry out programmes of research and publication. Many of these people advocate for flying-foxes and many also volunteer time and energy well beyond the demands of academic life. Like many others, they often use the word bats when discussing flying-foxes. Some people do this out of habit, others to try to combat the widespread prejudice against microbats by including all flying mammals under the single term. Others prefer to distinguish between Pteropids and other species and genuses. For clarity, I prefer the term flying-fox, using it to refer only to the Pteropids.

Invariably, volunteers are people who were seized by fascination and commitment and became determined to learn and do more. Tim Pearson, academic scientist, carer and advocate is one of the outstanding figures in the east coast region. He is a classic 'tall, dark and handsome' fellow: long, lean, intense, articulate as a teacher and immensely passionate, at times fierce, in defence of flying-foxes. And yet he occasionally flashes a delightfully sweet smile and he handles young and vulnerable creatures with great gentleness and patience. Tim told of his monitoring project in Sydney's Royal Botanic Garden (SRBG). Our conversation took place in the Garden while Tim was doing his identifying and counting:

> Purely by chance I was in here one weekend and I noticed a couple of Blacks up there. I asked around and no one had known about them. Someone suggested 'why don't you keep an eye on them', so that was the start of a four-and-a half-year obsession. I've been monitoring the whole colony and how the Blacks have been integrating in it.[17]

Tim's monitoring project was forced to change dramatically when the Royal Botanic Garden received permission to proceed with its plan to expel the flying-foxes. The SRBG management wanted to protect some rare and exotic trees that were being damaged by flying-foxes. Tim opposed the expulsion because he believed that other options for protection were both feasible and preferable. In his vigorous words: 'Disturbing one of the main maternity colonies of a threatened species that is a keystone species and a key pollinator of the east coast gum trees doesn't necessarily seem to be a sensible way to protect the trees.'[18] Having opposed the expulsion from the start out of concern for the flying-foxes and commitment to new forms of multispecies conviviality, Tim brought his monitoring skills to the subsequent expulsion that brought terror and trauma into the lives of thousands of flying-foxes (Chapter 7).

Camp life

Flying-foxes use their fingers for holding and touching, and indeed the hand/paw that became a wing is marvellously adaptable. The membrane of the wing produces its own waterproofing and can be wrapped around the body to make a raincoat; it is elastic and

self-healing and is covered with microscopic hairs that give information on wind pressure during flight. Blood vessels in the membrane can be regulated to retain heat or to allow it to dissipate.
When a female is giving birth, which she does while hanging upside
down, she uses her wings as a cradle and to help the newborn find
her nipple. Wings spread wide are a threat gesture ('big winging'
in local parlance); wrapped around a baby they keep the little one
warm and protected. The thumb is used for grooming and climbing,
and for holding on when a flying-fox has to turn head up to urinate
or defecate. Feet are also very active: they are the main organs for
grooming. The claws are prominent and are equipped with a lock-
ing mechanism so that the creatures will not lose their grip during
the long hours spent upside down.[19] Only in this way is it possible
for flying-foxes to carry on their social life in the gregarious camps
where they sleep, play, fight, have sex, give birth and nurse babies,
all while hanging head-down.

Females give birth to only one baby each year. Babies are born
with their eyes open, and their mother carefully moves them into
position so that they can latch on to her nipple. The fur on the
baby's chest is thin, allowing the little one to absorb warmth from
its mother. For a few weeks the baby hangs on to its mother's nipple
with its teeth, clutching her fur with its claws, while she flies out at
night for food. Later, though, babies are left behind in crèches in
the centre of camp. The mothers return at dawn, flying round and
round until they locate their own baby, and re-attach the baby to
the nipple. Once the babies have grown into adolescents or young
adults, they leave their mothers and move into their own age-mate
groups, forming lively, noisy crowds. Senior males are said to take
them out at night, teaching them flying techniques, and showing
them how to forage and to find their way home again.[20] Hall and
Richards describe these clumsy youngsters:

> They do not have the purposeful direction of the adults, and are
> reminiscent of a group of school kids going home from school
> and exploring their environment. Progress is slow as they carry

out aerial bombs on each other, explore vegetation and duck from imaginary predators. It is probable that these groups do not initially go far from the camp, and that the trips serve as navigational training.[21]

While flying-foxes prefer large groups, most urban camps are only a few thousand strong, perhaps tens of thousands in some places at some times. They hang out together all day, sleeping but also doing a lot of socialising. With at least thirty different vocal calls, all of which are audible to humans, they are, from a human point of view, very noisy folk.[22] Tim Pearson explained the situation at Sydney's Gordon camp, a large area of bush with a small stream in a steep-sided valley in the Sydney residential area known as Gordon:

> The residents and the bats live in a mostly uneasy truce. Some people love them. Some people think it's an absolute wonder and a privilege having the bats there. But even some of the people who think it's a wonder and a privilege, when the colony swells – like, February of last year – to seventy thousand bats and they're all juveniles and, like, all night and all day it's really noisy – even the most ardent bat fan can get a bit short-tempered, and I can't really blame them because, basically, you have to sleep with ear plugs in.[23]

Along with being active and noisy, flying-foxes are also disarmingly delightful to observe. One of the great treats at Sydney's Royal Botanic Garden used to be watching the flying-foxes 'belly dip' in the lagoon in the late afternoon. Tim described this while we watched and filmed individuals leaving their place in the tall trees and swooping down to the lagoon:

> They skim over the water, get the water on their belly, then fly up to a tree and lick the water off and have a wash in it. Occasionally you'll see some belly dipping over the other side in the salt water. We think that's because they need a bit of salt . . . You can tell the

Figure 2.1 Flying-fox belly dip
Source: © Nick Edards.

ones that are the older bats that really know what they're doing. They usually make one circle around, approach beautifully, flap in and just glide with enough speed, do it perfectly, and fly off. Some of them have a few goes at it, they don't quite have the courage to commit. We've occasionally seen very young ones miscalculate and just totally face-plant . . . One very hot day we saw one juvenile come in and misjudge, and whoomph! His little head popped up and he swam ashore and climbed ashore very wet and obviously a lot cooler.[24]

Given their desire to camp together in very large numbers, both flying-foxes and trees do best where the area is large. Tim monitored a number of Sydney camps, going out with a tracking device that enabled him to record the presence and approximate location of flying-foxes that had been fitted up with a radio collar. He explained that the Gordon camp is a particularly excellent urban camp because of its size:

It's the closest thing to a natural camp in the Sydney region. It's in a valley. It's in semi-rainforest. There's dense mid-storey and ground cover that locks the moisture in . . . The flying-foxes move around a lot within the camp areas. No trees become too stressed. A lot depends on undergrowth and microclimate as to where they camp.[25]

We were there in winter and the flying-foxes had moved away from the valley to camp on the hillsides. Tim conjectured that the valley floor had become too damp and cold. In contrast, summer heat-waves draw them back down: 'You go into Gordon on a heatwave day, and the bats have climbed down – when it gets really hot, they're hanging this high [a metre or so] from the creek. They climb down and down and down.'[26]

Camp populations have distinctive patterns. In a study of a camp near Gosford, north of Sydney, the academic scientist Kerryn Parry-Jones and her colleagues found two patterns of inhabitation operating at once: the local flying-foxes who were always present in the mating season and who numbered about 20–26,000; and the transients who were responding to a local abundance, and whose numbers increased the camp population to around 80,000. The other pattern they documented was the seasonal one, with all the flying-foxes leaving during the winter months (July, August, September), returning shortly after having given birth to establish a crèche, raise the young and mark out mating territories.[27]

Where the tongue meets the tree

There is a fantastic love story of the co-evolution of flying-foxes and Myrtaceous trees and shrubs. Flying-fox life works between two great imperatives: travel in search of food and sociality, and strong site fidelity. Both imperatives are oriented towards and coordinated with the flowering of their preferred foods.

The most beloved of foods for flying-foxes is the nectar and pollen of Myrtaceae trees and shrubs. Flying-foxes live, or lived, by chasing blossoms of Eucalypts and Melaleucas by preference, with Protaceae (such as the *Grevillea robusta* in my backyard) also offering important sustenance.[28] In Australia, Eucalypt blossoming takes place sequentially; flying-foxes are readily able to know when trees start to bloom hundreds of kilometres away from where they are camping, and so they fly off to find the nectar; scientists do not know how they do this.[29] Parry-Jones reported on a camp of about 80,000 individuals in New South Wales where on one night in June 1989 almost the whole mob left the camp, flying away to destinations unknown.[30] Where did they go? We don't know. Why or how did they all decide to leave on the one night? We don't know.

Many of these trees and shrubs produce huge clusters of light-coloured showy flowers that are perfectly adapted to be visible at night. Furthermore, many of them actually produce their nectar and pollen at night so that their peak of desirability coincides with the night-time flyout of their main pollinators, leaving the leftovers for the bees and birds to consume during the day.[31] Moreover, the trees themselves are more receptive to pollination at night.[32]

The exact shape of a Megachiropteran tongue is adapted, species by species, to the food on which each particular species depends.[33] Flying-foxes travel up to fifty kilometres per night getting food, and their foraging patterns get them moving from tree to tree across wide areas. They are the primary pollinators for numerous species, including rainforest species for whom they are also seed dispersers.[34] Flying-foxes, for their part, are highly attuned to smell, and have excellent night vision that is especially attentive to light colours. Tim summed up the information on flying-fox senses, hoping in part to overcome the misperception that they are blind: 'Their daytime vision is as good as ours, and their night-time vision is generally reckoned to be about twenty times as good. Their sense of smell is acute. Really amazing. And their hearing is quite incredible too.'[35]

Because these trees flower sequentially, 'myrtaceous forests and woodlands provide a constant food supply throughout the year for these animals'.[36] At the same time, many Eucalypts actually require pollen exchanges with more distant trees, and thus are best pollinated by the highly mobile flying-foxes.[37] Habitat fragmentation now is isolating more and more patches of trees, threatening to trap them within a small genetic community, and yet in this time of climate change plants need to have the capacity to adapt quickly.[38] Australian ecologist Tim Low explains: 'More pollen exchanged means more varied offspring, improving the chances that some of them will have the features that suit the future.'[39] Widespread pollination increases the chances for adaptive changes, and flying-foxes do exactly this long-range work.

Many of the plants preferred by flying-foxes are notoriously irregular in their flowering,[40] so flying-foxes' capacity to travel so far, and their mysterious ability to be immediately and coherently responsive, ensures that they are superbly adapted to the patchy distribution of Australian flora, and to the boom and bust pulses of Australia's El Niño-influenced abundance; it is probable that the well-being of both plants and flying-foxes is linked to their ability to respond opportunistically to the unpredictabilities of El Niño events.[41]

With their long-distance pollination and seed dispersal, flying-foxes are keystone species in the survival of the plants and ecosystems that depend on them. Foraging in rainforests, Melaleuca swamps, mangroves and swathes of Banksia heath, along with the dry sclerophyll woodlands, the nightly work of flying-foxes is mutualist and life-enhancing. One of the preferred Eucalypts along the east coast is *Corymbia maculata*, commonly known as spotted gum. These trees flower quite irregularly, with intervals of between four and ten years. When they do flower, however, they continue for as long as six months, blossoming sequentially from north to south.[42] When they flower, flying-foxes follow.

What remains of Australian native forests now is only a minute fraction of what existed prior to British settlement. In the area where Parry-Jones and Augee conducted their research, spring, which is the birthing time, was the hardest for flying-foxes because European settlers had removed almost all the trees that flower in that season.[43] Rainforests provided an alternative, since some of the trees, figs in particular, produce fruit all year round. However, over 95 per cent of the rainforest has been cleared. The devastation of other Australian ecosystems is well documented in relation to flying-foxes. In south-east Queensland, for example, 'approximately two-thirds of formerly continuous bushland cover had been cleared by the early 1990s' including vast tracts of *Melaleuca quinquinervia* (paperbark) 'which is a primary food resource for nectar feeders in the winter months'.[44] Furthermore, 'many scattered forest remnants are now located on nutrient-poor soils, on which trees have been shown to have low nutrient productivity and reduced resource yield'.[45]

Land clearing is a socially contentious issue, and in recent years political pressure from both primary producers and developers has induced NSW and Queensland governments to ease restrictions, with the predictable result that ever more land is being cleared. The challenge to ecologically concerned activists intent on saving native forests and bush is ongoing, with more losses than gains, and the impacts on native flora and fauna are pushing ever more species towards the edge of extinction. The debates about extinction are more political than ecological, and much depends on the willingness of the public to force politicians to protect species and habitats. As is to be expected, uncontentious charismatic animals such as koalas are key figures around which public opinion is mobilised. Tim Pearson offered a bruising dose of realism: 'If we can't even save forests for koalas, then we won't be able to save them for any other species.'[46]

Cities, towns and farms

Loss of habitat and loss of preferred foods leads flying-foxes ever further into areas of conflict. Throughout the east coast, in particular, when their preferred foods are not available, they turn to orchards for fruit. These days many orchardists put netting over their fruit. Some, however, continue to shoot. One of the most curious aspects of this failure of species consideration in the state of New South Wales is that the same organisation charged with protecting the creatures because they are listed as endangered, the National Parks and Wildlife Service of the Department of Environment, Climate Change and Water, is also charged with issuing licenses to shoot them. It seems certain that a great deal of unauthorised shooting also goes on. Tim said that 'at the Central Coast there's some orchardists who say they don't want the netting and say they'll never net because they enjoy their barbeques on Saturday night, when they get their mates around and blast bats out of the sky, and they're going to keep on doing that, by God, no matter what the law says.'[47]

Perils of the bush include not only habitat loss, and not only men with guns, but also life-threatening human-made products. A flying-fox tangled in barbed wire quickly becomes injured and distressed. In their efforts to free themselves they become ever more tangled and injured, and it takes a skilled rescue person to free them. Small changes to how fences are wired up make huge differences for flying-foxes, and education programmes are part of ongoing advocacy.

It is important to remember that long before there were European settlements, flying-foxes and Aboriginal people shared country; people cared for country, flying-foxes pollinated trees and shrubs, people ate flying-foxes, and through a variety of mutualisms everyone flourished. During the long millennia before humans came to Australia, flying-foxes ranged without interactions with humans, and they thrived. Of course, there were always perils in the bush. In the tropics, crocodiles are one of the main predators, often snapping them up when they belly dip. Pythons are another regular predator, and so too are white-bellied sea eagles, powerful owls and a range of other birds.[48]

When humans disturb flying-fox camps in order to drive them out, individuals are exposed to predators in ways that wouldn't formerly have been at all usual. One Northern Territory stockman described a dispersal that sent flying-foxes into the sky in broad daylight where sea eagles mustered them into groups and herded them away to their death.[49]

Cities and towns, orchards and paddocks with barbed wire, all these objects of human inhabitation moved into flying-fox sites and country. Hazards abound, and there are also new opportunities. The bush has become more perilous for flying-foxes, but towns and cities have become more welcoming from a flying-fox perspective. Given their non-negotiable needs for water, food and stands of tall trees, Sydney is a good example of urban opportunities, offering generous areas of municipal parks, providing extensive waterways, street lights, artificial watering in public spaces, and ready access to large National Parks and other protected areas within flying distance of the city. In addition, popular interest in planting Australian native plants along streets, in parks and in backyards has meant that there is more native food than ever in cities. Many householders plant fruit trees in their backyard, and the fruit season is an attractive time for flying-foxes. So when flying-foxes find a place where there is food on a regular basis, where the microclimate offers humidity and shelter from the heat, where they can cool themselves in summer and warm themselves in winter, and where navigation is facilitated (often by the presence of street lights), they are inclined to take that place very seriously.

Inevitably, though, perils also are part of the story. Cats and dogs may grab flying-foxes should they come close to the ground. Until the widespread use of unleaded petrol, lead poisoning had an impact. There are collisions with obstacles such as motor vehicles and airplanes, and there is electrocution in electrical wires. Entanglement in barbed wire and garden netting add to the perils. Pesticides and exposure to new bacteria may also be having an impact.[50] Human anxiety about the proximity of flying-foxes is fuelled in part by the fact that these creatures have recently been found to carry two viruses that are lethal: Hendra virus and Australian bat lyssavirus. Hendra is now preventable, and lyssavirus, a close relative of rabies, can be prevented by rabies vaccine. It is lethal only if untreated. Both diseases have recently started to spill over from flying-foxes to humans, and the best protective measure is to ensure that only people who are vaccinated handle flying-foxes. As far as is known, three people have died of lyssavirus.[51]

It is tempting to think of urban flying-foxes as refugees, and undoubtedly many of them have experienced life in that vulnerable manner, but that is not the whole story.[52] We need to remember that the places where humans settle become part of flying-fox range; new opportunities and adventures arise for these mobile, curious, adaptable creatures whose home places have become cities and towns. We could remember, as well, that refugees (human and nonhuman) arrive with gifts as well as needs. Flying-foxes do the work of pollination and seed dispersal in cities as well as the bush. And for many of us urban humans, a city with flying-foxes is an utter delight. As the journalist James Woodford wrote, 'watching bats silhouetted against the stars is one of the greatest, but little known, pleasures of life'.[53]

* * *

Urban and suburban camps are becoming more permanent. The once nomadic flying-foxes are changing their ways.[54] These new camps still require nightly flyouts for foraging, and the numbers fluctuate considerably, but some flying-foxes stay year-round. In this way they are exchanging the wide-ranging reliance on episodic abundance in the bush for the food security of locally occurring plants, many of which are not preferred, but whose regularity is significant. One of the effects of urban life is that camps tend to be smaller in area, and smaller in numbers of individuals. When flying-foxes were regularly on the move their impact on the trees in which they camped was limited by the fact that when they left the trees had a chance to recover. With more permanent settlement, especially in small stands of trees, recovery is difficult, perhaps impossible.

The evidence of damaged trees is widespread, and perhaps most worrisome to humans in the context of botanic gardens where the statutory focus, and indeed the passion of plant-lovers, is to protect the flora. For example: Melbourne's Royal Botanic Gardens. Melbourne is a southern city, and once was outside the range of flying-foxes. Climate change and the unique properties of urban climates have meant that Melbourne has become habitable. Flying-foxes first came to the Royal Botanic Gardens in about 1981, occupying a small area known as Fern Tree Gully between 1985 and 2003. At first they were welcomed as a novelty, but as their continued presence began to damage the area, the decision was made to 'exclude' them.[55] Methods of removal began with a 'cull', a sneaky term which seems to disguise the fact of slaughter. Public outrage stopped the deathwork, and in the year 2003 other measures were tried.

> Substantial resources, time, and effort were committed to moving the colony. Attempts to lure the bats to roost in a pre-prepared, alternative site (complete with bat enclosures, decoy bats, audio recordings of flying fox camp sounds, leaf litter from the RBG, misting devises, and the daily provision of fruit salad) proved to be unsuccessful. However, dispersal and directed 'herding' of the bats, using a combination of sounds that had been found to disturb the animals eventually had the desired effect.[56]

The ecologist Tim Low discusses what he calls a disinformation campaign designed first to justify the killing and later to justify the expulsion. His main points include the fact that a small area of the Gardens was discussed as if it were a large landmass. The threat of flying-foxes was stated in greatly exaggerated terms: they would destroy the Gardens, they would destroy the scientific work being done there, and they would destroy the civic amenities Melbournians enjoyed there. He points out as well that 'the media, instead of exposing this as preposterous nonsense, played along with it'. Low scrupulously offers ample evidence for his analysis.[57] The cost of this removal programme was said to be 1.7 million dollars.[58]

Tim Pearson offered an abbreviated account of this story from his vigorous perspective:

> Melbourne! In Melbourne they start roosting in the Botanic Gardens, in a thing called the Gully. Lovely bat habitat. Six years

ago, seven years ago, they culled them. Started shooting the bats. Threatened species. This is what we do in Australia. We're bloody geniuses! Threatened species! Keystone pollinator, so let's just blast them out of the sky!

Anyway, so, Melbourne decides they're going to relocate the bats out of there. They look at the situation. At that point, Melbourne is at the extreme south of the Grey-headed flying-fox range. It's a single colony at the tail-end of a species' range. Everywhere else has multiple colonies dotted around. Melbourne hasn't got a network of other colonies around it. It's this single, isolated colony. They decide they're going to drive the bats up river to a new area. They get this area scoped out, choose it as perfect bat habitat, they build a cage, put captive animals in there – trapped bats from the colony, put captive animals in the cage to attract the other bats, then they start the noise dispersal.

The bats scatter all over Melbourne into parks and stuff. People ring up the rangers. They come and chase them on, the bats go in the opposite direction. They end up in a place called Yarra Bend. It's three and a half kilometres from the Botanic Gardens, and there was a green corridor that they went along . . . Hallelujah, it's a great success! The bats have gone to Yarra Bend. Checking the second and third year after they were relocated, they're breeding perfectly in Yarra Bend. Notice anything wrong with that? . . . [First year:] No births. Virtually no births. They knocked out a year's breeding. So, second or third year, they're breeding in Yarra Bend. It's a perfect success. The last two years, six thousand bats have died from heat stress in the Yarra Bend colony, and in the previous ten years no bats died in the Gully in the Botanic Gardens . . . now [they've got] this massive vegetation program. Chuck another million bucks at it! The colony's fragmented, a new colony's formed in Geelong and two other places [Bendigo and Doveton]. So this is a successful relocation![59]

For a species in decline, the loss of a new generation is significant. At the same time, some carers were also inclined to consider the

ongoing problems of flying-foxes in concentrated areas. In the view
of Jenny Mclean: 'I think they had 50,000 down there. There's no
way you could have them in the gardens. I think it was the right
thing to do, and they did it the best they could . . . It wasn't sustainable to have them. But nowhere is as good as the gardens were. . . .'[60]

After the expulsion, then came the heat stress. During a subsequent heatwave, deaths at Yarra Bend brought out volunteers
during the hottest days of the year.[61] They set up pumps and sprayed
the flying-foxes with mist to help cool them down. It was one of the
first large heat death events to be well documented, and it was
immensely traumatic for humans and flying-foxes alike. In Tim
Pearson's words:

> Heat events are horrible. You're surrounded by dying animals
> who you can't help. You're spraying them in the trees, trying to
> help them, you're surrounded by death. They literally clump and
> then just slide down the trees, and you've got a melted candle of
> dead bats, so you end up rummaging through piles of dead bats
> hoping one moves because it means it's alive, and you can possi-
> bly save it. It's horrible, it really is.

The zero-tolerance approach when it comes to human interests
constitutes a refusal of multispecies mutuality based on flexibility
and consideration. Again and again this approach results in trauma
to flying-foxes. Tim takes it hard; indeed, everyone who cares and is
involved takes it hard. 'I kind of apologise to every bat I rescue', Tim
Pearson explained. 'I have to apologise for my species!'

Excursion to Port Keats

Dusk, the transitional spell between day and night, is a stimulus
for action. Most of us diurnal creatures are keen to get ourselves to
safe shelter, or at least close to bright lights, but the nocturnals are
called to go forth and be active. Between bright day and darkened
night, there is a brief interlude of dusky visibility that is the best

time for seeing flying-foxes in their winged glory. The most spectac-
ular flyout I have seen took place in 2014 when a huge mob of Little
Reds congregated in the vast mangrove swamps near Wadeye, an
Aboriginal community in the Northern Territory formerly known
as Port Keats. They were drawn, as far as is known, to a tremendous
Eucalypt flowering not far inland in the savannah.

Darrell Lewis, archaeologist and historian, had been in Wadeye
and he sent me a text message saying there was a flyout surpassing
anything he'd ever seen. 'Everyone was talking about it', he said,
'whitefellas and Aborigines. Come dusk, they'd set aside whatever
they were doing, drawn to the sheer amazement of the drama they
knew was coming.'

Wadeye is south-west of the city of Darwin by several hundred
kilometres, and if I could get myself to Darwin, Darrell would take
me the rest of the way. I arrived a week later, hoping like mad
that the flying-foxes hadn't all decided to go somewhere else in the
meantime. The country was touched with gold by the late afternoon sun as Darrell and I drove the last kilometres of gravel road.
We saw smokes on the horizon where people were burning the
country. Closer at hand, the fires had come and gone, and slanting
sunshine lit up the black trunks and brilliant green branches of
new growth. We raced past ant hills with fluted tops, and past
black and green cycad groves where new fronds formed shapes like
gleaming wine glasses resting on top of tall black stems. We slowed
down through paperbark swamps, and crossed small, clear running
creeks. Through hot scrub and entrancing flats, we chased the lure
of flying-foxes.

Once we got to Wadeye we positioned the video camera, almost
holding our breath as we waited. I filmed the sunset, and still we
waited. The sky became a deeper mauve, the blue vault grew darker,
and still we waited, watching a line where the mangroves met the
sky.

When the first creatures appeared, it was as if a treasure box
had sprung open. The horizon by then was almost black, and suddenly it started to fragment. Flying-foxes arose in their thousands

separating from the trees and from each other, taking flight and heading off towards distant places. Some travelled low towards some faraway blossoms, while others spread out over our heads. The sky was thick with them, and we could hear their wings fanning the air. From time to time, one would turn and go back, but the vast majority kept going in the direction of their night-time feast.

It is impossible to gauge the number of individuals in a flyout unless one is an expert. Large numbers are by no means impossible, and camps of more than a million are reliably documented for the coastal swamps of the tropics.[62] Even in the state of New South Wales where the numbers are in decline, there were reports of camps of 200,000 individuals as recently as the 1980s. I was reminded of how the naturalist Francis Ratcliffe described a cloud of Grey-headed and Little Red flying-foxes in southern Queensland back in the 1920s. At that time the populations were already in decline because settlers had been killing them in large numbers. Here is his description of a flyout:

> The scrub by that time was belching forth foxes. They rose up in thousands, circled once or twice, and then joined the southbound stream. In three or four minutes a column of the beasts about a hundred yards wide was stretched away across the sky as far as I could see.[63]

One present-day ecologist gave the following reckoning:

> From a very rough census I estimate some of the flocks which congregate together for shelter in the daytime number hundreds of thousands. Not so long ago a few must have crossed the million mark.[64]

The awe I experienced at seeing the Little Reds who made up the Wadeye mob was tinged with my knowledge that in many parts of Australia flying-foxes are reviled and persecuted. Even as I treasured every moment of the flyout, I couldn't help but hold in the back of my mind the sense that someday this sight might no longer be possible. For the moment, though, the experience was exhilarating.[65]

Wadeye has its problems, as do many communities, but it also has its strengths. One of the great sources of power arises directly from Aboriginal culture: an enthusiasm for participating in the joy of life as it arises when people recognise the awesome wonder of the lives of others. One of Darrell's friends told him that the previous night the flying-foxes had varied their track slightly and flown directly over the community. The children were out playing, getting the most out of the last light of the day when the flying-foxes flew over. The kids stopped their play, and they cheered!

The great flyout grabbed our attention and exhilarated us; it brought us into splendour. We realised, yet again, that this is how life is meant to be: country that is well cared for; animals free to lead their own lives of purpose and beauty. And in the midst of this glory, humans who respond in kind: cheering children; watchful, awestruck adults.

3   Arts of Care

Death by heat

The year 2013 was the hottest ever recorded in Australia. According to one report, '2013 will go down as the year that registered Australia's hottest day, month, season, 12-month period – and, by December 31, the hottest calendar year'. This was the year that scientists at the Bureau of Meteorology decided to add two new colours to the temperature maps. Deep purple and pink joined the colour coding to indicate maximum temperatures of 50–54°C (122–129°F).[1] The trend continued into January of 2014 with a massive heatwave, and the result was one of the worst mammalian mass deaths in recorded history.

The connection between heat and death is this: when temperatures reach 40°C (104°F) flying-foxes start to suffer. If these temperatures continue, the suffering leads to organ failure and death. Basically, when temperatures reach 43°C (109°F) they 'start to melt from the inside out', as one scientist vividly described it. Heatwaves of this magnitude generally occur in January but may start much sooner. Most of the young are born in October; in January, many females are still lactating, and babies may still be dependent on their mothers. Both mothers and babies are extremely vulnerable.

In the 2013 heatwaves in Sydney, for example, 15,000 individuals died over the course of a few days. The impassioned carer Storm Stanford pointed towards the amplifying magnitude of the disaster: 'Ninety-nine per cent of the animals that are dying are juveniles. We are losing a significant proportion of the next breeding generation.'[2]

Australia is familiar with heat; the continent is not only hot and dry but has experienced long-term slow drying.[3] Flying-foxes adapted for millennia by seeking areas where there was ample water, with humidity and/or breeze to mitigate the effects of heat. Certainly, there would have been heat stress deaths in the past, amongst animals and plants, and humans too. In written recorded history an officer of the First Fleet, Watkin Tench, 'wrote that flying foxes and birds fell dead from the sky during the searing summer of 1789, when north west winds blew across the city and turned it into an oven'.[4] Under the emerging pattern of extreme weather events, the challenges for flying-foxes are dire because the combined impacts of habitat loss and persecution are amplified by climate change. As humans take over areas of former flying-fox camps there are fewer and fewer options for weathering the heat. At the same time, global warming is now happening rapidly in Australia. 'Weather on steroids', is how various science writers put it.[5]

The January 2014 heatwave was relentless, and the suffering went on for days as death worked its way through flying-fox camps in eastern Australia. Queensland seemed to have experienced the worst of it. As the heat progressed, some flying-foxes fell out of the trees or crawled down and died on the ground, suffocating others who were beneath them. Some left this life with their claws still locked on to a branch. Babies clung to dead mothers, and struggling mothers held dead babies. Wildlife carers worked their hearts out trying to save lives. The high temperatures were stressful for humans, too, of course. Rescue is hard work physically as well as emotionally, and the carers were almost overwhelmed with the magnitude of it all. In a massive effort to deal with the numbers, orphans were airlifted to carers as far afield as Sydney.

I have not located a final mortality figure. The Royal Society for the Prevention of Cruelty to Animals offered a figure of 100,000, based on twenty-five Queensland camps.[6] Since then, lower figures have been suggested, but the fact is that no one knows with certainty. Three academic scientists who collaborated to describe the event documented 45,000 flying-fox deaths on a single day in camps in south-east Queensland; the deaths continued for many days moving through New South Wales and on into Victoria and South Australia, so the numbers must be high.[7] It was the greatest mammalian mass-death event to be caused by the new patterns of extreme heat that are Australia's experience of global climate change, and it was the first of many. Who will live and who will die have become questions of temperature, refuge, assistance, individual condition and luck.

The work of compiling a death count is both traumatic and difficult, as Tim Pearson explained in the context of the Yarra Bend event:

> It's natural for anyone involved to overestimate. The numbers we have . . . have been obtained by collecting all the dead bodies they can find from the camps. So, I'd say they're close to accurate, but I can certainly understand someone who's on the ground thinking that there's got to be more than, like, in February last year, about five thousand or so being on the ground that day. Five thousand dead bodies is a lot. It would look like tens of thousands. It would look huge. It's heartbreaking, too. I've been through one heat event. I don't want to go through another.[8]

That conversation took place in 2010. Australia has continued to experience strong heatwaves and flying-foxes' deaths are in the news every summer. Again and again, carers do the best they can. Much of the clean-up is not exposed to public scrutiny, perhaps because it is so disturbing. People in hazmat suits rake and shovel dead bodies into piles and then load them into wheelie bins for onward disposal at designated dump areas. It takes time, and the dead bodies attract flies and maggots; many local residents under-standably complain about the smell of all that death. And yet, this close encounter with mass death is just one of many stories of the Anthropocene. Flying-foxes, like us, are creatures who suffer and die, and they are members of species that may yet be lost in the rolling apocalypse of change that is becoming cataclysmic. Their vulnerability is the visible crest of a wave that is bearing down on earthlings at a rate and with effects that we cannot yet imagine. Surely it is time to start.

* * *

During the January heatwave in Queensland, carers had been stretched almost beyond endurance. I waited until winter to visit, hoping that by then the trauma would have receded and that some of the people would feel able to talk with me about their experience in the midst of mass death. It quickly became clear that the trauma was still fresh in people's hearts and minds. According to Denise Wade, one the carers who had been deeply involved in the rescue: 'I know a lot of people who attended that heat event are still not over it. They're still very upset. The things that we saw. Yes, it was a horrible, horrible, thing.'[9]

Denise has received considerable veterinarian training over the years and is a flying-fox carer of renown. Blonde, soft-spoken, calm and remarkably unflappable, she is immensely capable. Her hands seem almost to have a life of their own as she picks up an injured flying-fox; examining, injecting and wrapping it in small blankets or towels, along with feeding, cuddling, massaging. She calls the flying-foxes 'bats' as is the custom in her area, and she choked up a bit as she described her experiences:

I'm a very emotional person, so when the phone rang and they said 'the colonies have gone down', I knew what we would see. When we got there we just got ourselves organised, and I said to my team 'leave the dead, just go for the living'. So they were just sorting through bodies, pulling out the living. It was a terribly hot day . . . We just brought everybody in who was still alive and gave them IP [intraperitoneal] fluids and put them straight into wet towels. We went back for six days. It was horrific. They were swinging in the trees . . . It was a sea of dead bodies just swinging in the breeze and there were bats all over the ground, dead, and there were babies screaming. It was just very distressing . . .

But it was wonderful to see all the volunteers pull together. They worked their guts out that day. Running in and pulling bats out to try to save them. Regent's Park – it's a remnant corridor between houses. Blacks, Greys and Reds were *in situ* at the time. . . .[10]

Here at our place it was 37°. The further west [inland] you went the worse it got because it was hotter, so it was a very regional issue . . . It was a dreadful period I hope we never see again, but I'm sure we will.

People [volunteers] were wonderful. We had people come from the coast. Offers for help. And the residents where we were, they were absolutely wonderful. They gave us iced water, they said 'come in, take whatever you need, come on in'. We were getting calls for days afterwards about orphans in trees. We'd go and get them out. On the Sunday we went back to Regent's Park and sorted through the dead, because there were still a few live ones in there. We went back months later and there were still bodies, remnants of bodies on the creek banks. It was a dreadful, terrible, terrible event.

Because it was January, the babies were a lot older. We had seven-week-old babies, an extra four hundred . . . We had to divvy them up and farm them out, and anybody who had any sort of facilities took some, even if only a couple. Everyone was full [to capacity]. Everybody worked very hard for quite some time. It was tiring. But we only lost a couple [in care], which was good. We pulled most of them through, which was great.

Denise explained that while they would have liked to keep records, there was just too much to do.

There were just thousands of them. It was horrible, something you didn't want to see again: and it's hard on us as volunteers. We have to provide all our own equipment, and we don't have a great deal of equipment to deal with that sort of stuff. We definitely need to be better prepared next time.

The really sad thing about Regent's Park where the bats were, you park up on the road and then you go down into the colony behind all the houses. Up on the road it wasn't too bad. There were breezes. But as you went down it got hotter and there were no breezes, so I think that's what did them in. Then they started coming out of the reserve and into people's backyards. And we had to be careful walking through the trees, because you've got adult male bats and a bit of a breeze, they'd drop. You had to really be careful. They were starting to drop. Their bodies were dropping out of trees. You don't want to cop a kilo of bat on your head! . . .

The birds were all half dead. The possums were lying on the branches, stretched out. It was terrible. It's really hard to see that sort of thing when you know the animals, how they suffer.

Humans were undertaking exceptional work, and so, too, were flying-foxes. According to experts, under everyday conditions the mother-baby bond is intensely focused and a mother who loses her baby does not adopt others. Female flying-foxes have two nipples but are rarely known to give birth to twins; only one baby is raised.[11]

In this time of mass death, and in the unusual condition of protected care, flying-foxes too responded to the calls of others, refusing to abandon the young and the vulnerable. Louise Saunders, a leading activist and carer, and former president of the organisation Bat Conservation & Rescue Queensland, wrote:

During these dark days it is heartening to witness some of the special behaviour our rehabbers [carers who engage in rehabilitation] witness whilst looking after our most maligned wildlife.

Trish awoke to a beautiful sight this morning when she went out to check on her charges.

Two nursing Black mothers had suddenly grown an extra lump under each wing. Yes, as well as the burden of their own babies, these gentle girls have adopted an 'extra' baby. Both of these babies were severely traumatised by the death of their mothers during the recent heat event and they sought and received solace from surrogate mothers.

Another rehabber went out to check at lunchtime and found an adult grey girl with a newly collared baby tucked up under her wing. She too has adopted a needy little orphan after the death of her own baby during the heat event.[12]

     * * *

Levinas tells us: 'For an ethical sensibility . . . the justification of the neighbour's pain is certainly the source of all immorality.'[13] Working at the front line of suffering, carers come to know first-hand this immorality that justifies away others' pain. Abandonment lies at the heart of it, and the positions are clear. There are those who respond, saying 'yes' to the calls of others who suffer; there are those who simply turn away; and there are those who actually side with death.

Louise Saunders witnessed the full range. She had been driven to horror by the fact that some people in a position to help refused to do so. I will return to the war against flying-foxes in later chapters. For now, I want to point out that prejudice against flying-foxes increased the death toll and also amplified the trauma for carers. More lives could have been saved if those in a position to help had been more responsive.

One of the main methods of intervening in heat stress is to cool flying-foxes by spraying them with mist. Volunteers work through terrific heat, spraying in the hope of averting trauma before it becomes too severe. In many areas, and at many times, local fire brigades have assisted. Such work falls under their responsibility for emergency rescue. But in one Queensland town Louise encountered a mayor who refused to call out the brigade. His justification was that as soon as Queensland reintroduced legal shooting of flying-foxes, his township would be in the front line of killing.[14] In 2014 the mayor was looking forward to future deathwork, and in the meantime he wasn't going to help save lives. This particular town was hard hit by the heatwave, and the flying-fox death toll was severe. In Louise's words, the mayor's decision was 'very callous. Very uncaring.' Members of her organisation reported the mayor 'for being so inhumane' and the council responded that it would investigate. Louise shouldered some responsibility, too, having been unprepared for such an event. Like Denise, she was becoming better prepared for next time. Overall, though, she was not optimistic: 'it's pretty tragic all around'. In an essay posted on her organisation's website, Louise posed the core questions confronting all of us in this time of loss:

Flying-foxes are complex and intelligent mammals and they feel pain, fear and joy. They are inquisitive, intuitive and sensitive.

Why does the human race choose to inflict such cruelty and suffering on such incredible native wildlife and why do governments justify their cruel policies knowing full well the cruelty and suffering they will cause?[15]

Everyday rescues

The need to offer care was intense during the heatwave, but flyingfoxes are injured all year round: they may become tangled on electricity wires and suffer severe burns; they may get caught on barbed wire fences, and then further injure themselves as they try to escape; they become entangled in fruit netting or are hit by cars; they are even (rarely) attacked by dogs or cats; and there is the occasional accident that is just part of life in a world of unpredictabilities. Such was the case for a flying-fox in long-term care named Jackson. I took a special liking to this little guy; his shy demeanour conveyed a quiet dignity. The information notes at the Calga Wildlife Centre where he lived, just north of Sydney, said that he had crashed into a building during strong winds. He was camped in Sydney's Royal

Botanic Garden, and he was very young. One evening when he flew out he got picked up by the wind, and was hurled into a wall. He was rescued, and although he could not be released because he was now blind in one eye, his quality of life was good and there was no reason to euthanase him. Jackson hit a wall, but it could as easily have been a tree. The elements of such a mishap are mainly bad luck: a sudden gust of wind, a solid object and an inexperienced youngster.

The need for rescue and care has seasonal periods of intense activity. Three of the four Pteropid species give birth at much the same time during September, October and November; Little Reds are an exception.[16] With all species, if the mother dies there will be an orphan to be raised, and as there are always casualties so there is a fostering season. Experienced carers start organising their supplies in advance, knowing that spring will bring orphans.

Another peak is the 'netting season'. People with backyard fruit trees may put up netting to keep away birds and flying-foxes. Some types of net and some ways of netting are more lethal than others. The worst-case scenario is that a flying-fox becomes entangled in the net. Householders may panic, but sooner or later they ring for help and a wildlife volunteer comes to cut the flying-fox out of the net and take it into care. It takes a firm, dextrous and gentle pair of hands to free a flying-fox from netting, and the same is true of other tangle events. I learned this the hard way on my first netting rescue. I was bitten, and the experienced rescue person who was mentoring me took over the job, completing it seemingly effortlessly. The family watched the process, and there was time to talk with them about better forms of netting and also to offer them some basic information about flying-foxes – their status as a protected species and their place in Australian ecosystems. This family was interested, unlike a previous netting rescue where we encountered a family who had settled recently in Australia. It seemed to be fear of the unknown rather than learned hatred that had induced them to try to kill the poor guy by bashing him with a rake before deciding to call for help. But to return to my own clumsy experience, we took the injured fellow to an experienced carer, Mandi Griffith. He calmed down completely when Mandi got her hands on him; she got the rest of the netting off and dressed the wounds. Fortunately, there were no breaks and his wing membrane seemed in good shape. Mandi said his mouth was awfully sore, and while it would take a bit of time, she thought he would be all right.

Some injuries are so severe that the individual must be sent to a veterinarian. Not all vets handle wildlife, and not all wildlife vets are willing to take on flying-foxes, in part because the need for vaccinations adds another layer of effort in lives that are already busy. According to Jenny Mclean, one of the outstanding carer/advocate/educators, 'there are whole towns in Australia where there is not a single vet that will let a bat come into its surgery, because they are as ignorant as the everyday person about these viruses. And if you can't have vets understand the situation, how can the general public understand?'[17] Nevertheless, flying-fox-friendly vets and advanced carers can enjoy good working relationships; vets oversee the use of barbiturates and assist with supplies of other drugs. Injured animals of many varied species are frequently brought to vets, and often they offer first-line care.

In all cases, whether it be netting, electric shock, barbed wire or other trauma, the first step for an advanced carer is to treat the creature for shock. Denise explained: 'They come in cold and in shock, so you have to get them warmed up . . . They're dying as they come in, so it's emergency triage. We get fluids into them and we warm them up. Most of them [survive], if they're viable.'

Tim Pearson has rescued individuals of many varied species, and in his experience flying-foxes are exceptional in their responses:

Yes, they're a wild animal. If one's caught in a net and you go to rescue it, it's going to scream and carry on and try and kill you. They're also bloody intelligent. Within a day or two – or less, depending on how you treat it – a flying-fox that has been trying to kill you and has potentially taken very large chunks out of you while you've been trying to rescue it, cut it off a barbed wire fence, or out of netting, or whatever situation it's in – is letting you scratch it on the tummy, hand-feed it . . . Now, that's a bit of a generalisation. Some of them are absolute sweethearts and some of them are right bastards, but all of them – of all the animals I've dealt with in wildlife rescue – and this seems to be shared by a lot of other people – they're the one [species] that actually seems to work out that, on some level, 'I don't know what's going on, but you don't appear to be hurting me, and what you're doing is making me feel better, and I now appear to be warm and safe and protected.' So much so that they'll put up with medication and many things until they feel they're better, and then you'll find them actually hanging by the door of the cage, or trying to get out.[18]

Much of this dedicated work takes place in people's homes. Denise lived in a quietly elegant home when I visited her in 2014. A contemporary courtyard-style place, the face it presents to the world is small and uninteresting, with a driveway leading to an attached garage, a few bland walls, and a Colorbond fence. Inside, it opens out to become a 'paradise' in the old sense of a walled garden. Here on the outskirts of Brisbane, in the subtropics of Queensland, the climate is mostly mild. Denise's paradise has rooms for study and for sleeping, and a large sunny open space where areas flow into each other and lead directly outdoors. The floors are cool tile, and in the laundry room she has everything she needs for her clinic. As I spent time with her at home I came to realise the truth of her statement that the flying-foxes had 'taken over just about every room in the house'.[19]

From the courtyard came the gentle sound of a fountain, and from the backyard we heard occasional chattering, squabbling flying-fox voices. Denise's husband Stuart built the enclosures for the convalescents. These luxury accommodations offer heaters to take the chill off winter, and plenty of tree branches to climb on and hang from. Denise and other volunteers cut kilos of fruit every day, and at night free living flying-foxes visit. The creatures at Denise's place are among the lucky ones. Their wounds were not impossibly serious, and they came into the hands of a supremely capable carer. She treated them as they were brought in and monitored their recovery. In due course they would go back to the bush, returning to their own version of paradise. It is a testament to their versatility that flying-foxes can both respond to human care and settle into cages, and also return to the wild. Pteropid paradise has no walls. It is a moving feast involving long-range travel, masses of blossoms and lively home camps in the company of thousands of others.

All flying-foxes are social animals and do best when they are touched and spoken to. Many carers regularly addressed the flying-foxes in their care as persons, and often referred to them that way as well. Every carer I met was deeply aware of issues of anthropomorphism, and none of them wanted to project human qualities inappropriately onto the creatures in their care. Living in such intimacy with members of a different species, sharing touch and food, homes, gardens and family, they were well aware of how different these others really are. Their ability to provide appropriate care depended on understanding flying-foxes in their own specificities both as individuals and as members of a species. By the same token, however, intimacy led them to understand that there are many similarities between humans and flying-foxes. As a matter of both respect and affection they gave their charges names and addressed them personally. The practice of giving a name to every individual that was not doomed enhanced their individuality from a human perspective. Conversations were clearly marked as relationships between distinct persons.

Denise was no exception. I mentioned that I had found it interesting the way so many carers addressed flying-foxes as persons. 'Yes, we do', she said. 'It's all right until you're out somewhere and you say, "that person was euthanased". People who don't know, think we're quite strange. We do tend to call them people. They are like little people. They have their own individual personalities. We do call them people.' She added, perhaps doubtfully, 'We shouldn't.' Then she addressed the flying-fox she was holding at the moment:

'Should we? Because you're not people, you're bats. Bats are much better than people.'

I held and cuddled Ruby, a Little Red whose mother had died after being hit by a car. Ruby was rescued as a newborn. She was lively and totally accustomed to human interaction. There was also a young Black who was terribly underweight. He had a systemic infection, having been tangled in netting on a custard apple tree, and left there for four days before anyone called for help. It was touch and go as to whether he would survive.

The day before my visit had been a tough one: 'We had our first miscarriage yesterday. A little Black girl came in. She'd been on a fence all day. She'd aborted her baby. She wasn't viable. That was the first one for us for the season. It's really sad.' Only a few carers are licensed to euthanase. Denise is one of them, and the law requires her carefully to account for every 'mil and part thereof' of the barbiturate that is used. 'I'm licensed', she explained. 'I'm licensed for euthanasia. If you give the sedative and then you euthanase, it's very peaceful, very, very peaceful.' During the summer months when many creatures are brought in off fruit netting, it gets hectic and tough:

> [I]t's pretty stressful in the summer months. We don't like it very much. We get a lot of non-fatal electrocutions where animals have been on power lines, especially in damp weather they don't die outright. They might get shockingly burnt, but they aren't dead ... We get a lot of those in summer ... It can be a little difficult at times.

Given the way Denise spoke to and about the bats in her care, I asked if she said a prayer or anything before she euthanased. 'No', she said, 'not so much. They usually get a lovely head massage and we talk to them. It's a very peaceful process, and I don't normally say anything specific to them, but we do talk to them and give them a head massage.'

Thick care and provisional knowledge

Maria Puig de la Bellacasa developed a pertinent analysis of the specificities that are integral to care, calling for a 'thick vision' of care. She quotes Donna Haraway's compelling words that 'nothing comes without its world'. This being so, we are called to consider others not as passive bodies but rather as thinking subjects inhabiting their own worlds of action and meaning. A world, in this thick account, includes the body, the self, the relevant environment, and the interweaving matrix that holds these elements in the dynamism of ongoing life. In the interfaces between species, thick care must be attentive to many particulars.[20] With flying-foxes there will always be much that we humans do not understand, but we are called to recognise that flying-foxes do inhabit their own worlds, and that our care must engage with enough elements of flying-fox life to ensure that both the body and the integrity of the individual's world are sustained.[21] A key point in Puig de la Bellacasa's analysis is the mode of knowledge Haraway calls 'thinking-with'. An example of this theoretical point is found in the touch and talking which carers offer along with medical first aid. Voice and touch recognise flying-foxes' special need for social interaction. This open-ended form of relating rejects ready-made categories and remains attentive to the particularity of creaturely worlds. In its wider implications, thinking-with 'enlarges our ontological and political sense of kinship and alliance'.[22]

An example of this theoretical point is found in the way Denise handled an injured flying-fox which was brought to her while I was visiting. A carer named Joie had rescued a young fellow from barbed wire. He was cold and in shock. Denise told Joie to bring him over, and she quietly set to work to heat up the humidicrib that occupied a large corner of the dining room, and to lay out the equipment she would need for first-line treatment. Joie brought in the injured Black, explaining that he had been entangled on only one line of barbed wire and had been there for just a few hours. Denise wrapped him and started a quick examination, talking to him all the while: 'Why'd you go and do that?' and 'Good boy, good boy' and 'Oh, yi, yi, yi.'

She found that although he had a broken tooth, his palate had not been damaged. Had that happened he would not have been able to survive. With some fluids and a sedative he started to calm down and drop off to sleep. Denise removed the broken tooth and tucked him up quietly in the humidicrib. 'You give him warmed fluids to help, his body temperature is extremely low. The priority is fluids and heat. He'll be very warm in the humidicrib. When he comes round, that's when we worry about drugs and stuff. The priority now is to get him out of trouble.' And to the injured flying-fox: 'Nearly there, darling. It's all right. It's all right darling, it's okay.'

* * *

As a general rule, thick care, as humans practise it, works with inherited knowledge, adding to it, perhaps revising it, certainly enhancing it.[23] In the context of Australian flying-foxes a fascinating problem arises: people who care for flying-foxes actually do not have a long history of intimate associations and received wisdom to draw upon. Aboriginal people know flying-foxes well as kin and Creation ancestors, but as far as I know, did not attempt to rescue, treat and return individuals to bush. Their attention was primarily focused on the well-being of their flying-fox kinfolk in their area, but not so much on individuals (discussed in Chapter 4). European humans have only lived in proximity to flying-foxes since colonisation, and practices of care are actually extremely recent. Caring well for flying-foxes involves knowledge that arises out of emerging, local, hands-on, practical modes of learning, developed through the accumulation of experience and information, and transmitted through direct mentoring along with some excellent DVDs, manuals and guidelines.[24] This type of knowledge is sometimes referred to as *techne*, distinguishing it from abstract universalising accounts that become the property of a specialised group of experts. *Techne* is embedded in local social, cultural, ecological and interspecies contexts. It involves actual practice, direct learning and sharing.

Received knowledge is often provisional and open to thinking-with; it is attentive to uncertainty and to engagement with the previously unknown.[25] At the same time, because it is not uniform, care procedures can become contentious; amongst carers whose commitments are so demanding, disagreements over 'best practice' serve as flash points through which numerous stresses are expressed. The significant point remains: carer knowledge is always becoming thicker, ready to deal with new needs as they arise.

* * *

We humans are endowed with capacity for empathy, and so too are many nonhumans including flying-foxes. Frans de Waal is a leading figure here. In his condensation of a lifetime of research, 'Putting the Altruism Back into Altruism: The Evolution of Empathy', he writes that 'empathy allows one to quickly and automatically relate to the emotional states of others'. His research shows that empathy is widespread across mammals and birds (and there is new research to show that something like empathy exists among plants as well[26]). As a scientist, de Waal asserts that there must be an evolutionary advantage to empathy, and he deduces that for social animals the capacity for empathy is integral to rearing new generations, and to sustaining social relations amongst adults.[27] This is certainly the case with Australian Pteropids.

Dominique Lestel calls the emerging recognition of nonhuman subjectivity 'the true scientific revolution of our time'. The import of this revolution is that '*the human being is no longer the sole subject in the universe*'.[28] As discussed (in Chapter 1), the term subjectivity refers to the capacity of an individual to understand itself as a thinking subject. So, what does it mean to say that some animals are subjects? Thom van Dooren offers a rich explanation in his study of ravens and crows, and the analysis is equally applicable to flying-foxes. As subjects these creatures are 'beings with their own understandings, their own modes of paying attention, of learning, remembering, becoming sensitive, and adapting understandings and behaviours'. He concludes: 'To be a subject is to inhabit an experiential world characterised by conscious experiences, by impressions, feelings, understandings and beliefs.' Subjects have histories, van Dooren reminds us; subjectivity arises 'inside long histories of biosocial intra-action and inheritance'.[29]

Subjectivity is integral to empathy; it is central to the practice of carers and is implied in much of what is written and known about flying-foxes. One consequence of subjectivity is that human engagement with an animal subject becomes an intersubjective encounter, relying on mutual recognition and the possibility of sharing 'similar conscious states'.[30] Carers speak of this capacity of shared experience as humaneness, perhaps implicitly contrasting their own intersubjectivity with the callous inhumaneness of those who inflict suffering. Not infrequently, they also use the term compassion. That face-to-face encounter with intense suffering and with death is often relentless. Feeling the suffering of others, carers suffer too. Louise spoke of the need to be patient and compassionate with other carers. In the depths of trauma, carers, too, needed care as they wept their hearts out 'over what they had witnessed'.[31] The work of care thus rests on empathy, but is actualised and amplified as people witness the dramas, joys and sufferings of others. Denise explained it this way: 'when you know them the way we do, you know how they suffer. They're so tough. That's the really horrible thing. They survive the most appalling injuries. And that gets to me too.'

Fostering

Many people in flying-fox regions take on the work of foster care each year. This care has a paradoxical aspect. Infant flying-foxes need social care: they need grooming and feeding, they need touch and voice, they need to feel they are the focus of the family. In short, they need attachment. Over the course of my research I heard a few sad stories of people who failed to offer enough social interaction and whose little charges did not survive. And yet flying-foxes are not pets; the aim of fostering is that individuals will return to the flying-fox world, and so they need to leave humans. It all happens rapidly from a human point of view: at one moment they are members of the family; the next moment they are gone forever. For their own safety it is best that they forget that they ever knew human beings.

Fostering has become an annual part of life for many individuals and families along the east coast. For many, this is the limit of the care-giving, but many of the more committed volunteers were recruited in this way. The social needs of babies mean that the orphan becomes part of the family, and other family members become part of the care environment both emotionally and through their direct and indirect contributions to the work. Tim Pearson explained how this took place in his home:

> Six or seven years ago my wife and I, after years of saying, 'We should get involved with wildlife rescue' finally went and joined Sydney Wildlife. At that stage in their basic training course, at the end of the two-day basic training course they had a session where different people would come and talk for ten minutes about the advanced courses you can do, and a woman brought in a flying-fox and said, 'You can work with bats!' and I went, 'You can work with bats? Are you serious?'
>
> Because you need vaccinations to work with bats, my wife said, 'Well, it's stupid, you having animals in the house and I can't help, so I'll get vaccinated as well.' She knew nothing about bats, but didn't care. The first year we raised a baby flying-fox pup between us, and she's been smitten ever since, as well. I mean, the background interest always being there, and then starting to work seriously with flying-foxes . . . I got more and more involved. As part of understanding the animals I was working with, I started studying them, first casually and then it just got more and more involved.[32]

In 2011 I met Naomi Roulston, a young woman in my neighbourhood who was fostering two orphans in her suburban bungalow. Naomi agreed to allow Natasha Fijn and me to accompany her and film her as she gave one of her young charges his final grooming and took him to the crèche where he would join other teenagers.[33] Naomi was a university student at the time, and she had turned her study into a nursery. Two infants had been offered to her for care and she accepted both, naming them Sabbaticus and Peebo. In the centre of the former study she had set up a clothes airing stand with towels attached to provide secluded areas. The babies could hang from the clothes lines, and when they started flying they flapped about the room exploring the curtains and practising their take-offs and landings. Along with the regular feeding and grooming, she cleaned the room frequently as flying-foxes do not become house trained.

Flying-foxes grow up in a one-to-one relationship with their mother, and part of the comfort of maternal care includes undivided maternal attention. This relationship is marked in part by the unique smells that baby and mother share. To solve the problem of how each infant could feel singularly cared for, Naomi had two long-sleeved shirts, each of which was worn only while caring for one or the other baby; neither shirt was washed for the duration of care. Feeding time was therefore complicated not only by needing two bottles and two lots of formula, but also by a quick change of shirt. Baby flying-foxes need to be fed every few hours, so this was a major commitment. I couldn't help but note and respect the fact that fostering was a particular effort for Naomi, as the time of need coincided with her final exams at university.

When we visited, Sabbaticus had already left home. We watched and chatted while Naomi weighed Peebo, and then took him outdoors to be in the fresh air and sunshine as she cleaned his wings and eyes with soft cotton, speaking to him and stroking him with full attention and great tenderness. She explained the grooming: 'This is what Mum would do. She would lick his face and everything. So it kind of makes him feel that I'm a real Mum. It's kind of reassuring, I guess.'[34]

She explained that Peebo's mother had been electrocuted on power lines, but the little baby attached to her nipple had survived. It is essential to get an orphaned baby into care as quickly as possible, and there is good cooperation between the electricity authorities and the wildlife agencies so that as soon as a flying-fox is reported to be tangled, especially during the birthing season, a carer arrives on the spot to wait for the electricity guys. It is usually possible to determine whether or not there is a baby, and so the carer waits, ready to rescue the baby if it falls off its mum, or to take charge of it once the electrical workers get it down. If by some rare chance the mother is still alive, she too will be part of the rescue effort.

The carer comes with supplies: blankets for wrapping the individuals, a rolled towel called a 'mumma' for the baby to hold onto somewhat as it would have held onto its own mother, liquids for rehydration, and a special infant formula so that nourishment can be resumed. It is touch and go as to whether such orphans will survive the trauma of their mother's death and their ordeal of waiting. After a few days of first-line care, orphans are handed on to dedicated orphan carers. Foster carers are astonishingly patient with their little charges, feeding the baby every few hours, day and night, cleaning and grooming them, keeping them warm, and giving them the touch and chatter without which they cannot thrive.

Baby flying-foxes are totally dependent on their mothers, and of necessity there comes a time when the relationship has to end. This happens at around the age of three months. In the bush, the juveniles cease to be dependent on their mother's milk; they leave their close proximity with her, spending their days and nights in crèche groups within the camp, hanging out together and learning the skills they will need for adult life.[35] Youngsters in human care indicate through their own actions that they're ready for separation, becoming less cuddly and more assertive, making it clear (to those who understand) that they are ready to move on. Peebo was by turns cuddly and aggressive, showing his readiness to leave by striking at Naomi's hair and engaging in similar distance-inducing gestures. Carers provide for this stage of life by designating a few long-term care facilities as crèches. Within the crèche enclosure are adults who know how to interact as flying-foxes. Here the young-sters learn flying-fox etiquette and gain the communicative and other skills they will need when they are released.

The crèche we went to is run by Mandi Griffith, a dedicated and extremely gifted carer as well as veterinary nurse, known to friends as the 'bat whisperer'. The whole of her backyard is given over to enclosures, treatment areas, and work areas where she and her group of volunteers cut enormous amounts of fruit for the crea-tures in their care. Pieces of fruit are strung on wires and attached to the top of the enclosure where individuals can munch at will. Many injured flying-foxes are brought to Mandi, either from rescue people or from veterinarians; she cares for them all, releasing many back into the bush. All the flying-foxes that come into Mandi's care are weighed and measured, given a number and an ID card, and banded temporarily so that they can be identified until Mandi gets to know them as individuals. Later, if they are to be released, they are banded more permanently for future identification.

Peebo's experience of entering the crèche was typical. He was delightfully curious but a bit intimidated by the new place and indi-viduals. Naomi left his special shirt in the enclosure as a security blanket to help him through the transition. A number of flying-foxes in the enclosure came to check out the newcomers (Peebo, Naomi, Natasha and me). One lively little fellow in care was named Thor. He had been injured by electricity, but was alert, curious and quick-moving. Thor came to investigate me, and having inspected my hair he started on my forehead. I felt his slightly raspy tongue on my eyebrows. It was a hot summer day in Sydney and I was sweat-ing. Suddenly I realised I had become part of an unexpected and awesome interface: 'He's after salt. I have become Thor's salt lick!'

Mandi's crèche works so well because it has an appropriate group of permanent residents. Appropriate means there is a bal-anced range of age and sex. These flying-foxes are integral to care and advocacy, but they are not arbitrarily kept in order to serve the needs of others. Rather, these are individuals whose wounds are so severe that they won't be able survive in the bush; at the same time their capacity to engage with others – both flying-foxes and humans – means that they are able to take up other roles in the interface between flying-foxes and humans. A great glimpse into the lives of enclosure-based flying-foxes is shown in a video made with Mandi and the flying-foxes in her care.[36]

Later in the year Peebo and the others who were ready to leave were taken to a soft release centre. Here newly free flying-foxes have an enclosure in which to live where they will feel safe at first and where small amounts of food are provided. The doors are all open, bush flying-foxes come by to check out the newcomers, the released flying-foxes go in and out, and they soon take up with others, learning to live without human assistance as they become young adults and join the shifting and gregarious mobs.

* * *

Not all flying-foxes make it back into the free-living bush life. However, some individuals in long-term care show an aptitude for meeting the public. Carers bring 'education bats' to schools, fêtes, community fairs and other similar venues where the public can get close to, and learn about, wildlife. A good educator is willing to be stared at, to have its wings stretched out to show the size and the leathery texture, to be part of a show that is both informative and, for most people, extremely engaging. Face-to-face encounters in safe and enjoyable circumstances are part of broader public education programmes targeting people who live near flying-foxes and who need to understand the means and possibilities for mutual co-existence. Research indicates that these education efforts have positive impacts.[37]

Education bats form a cohort of creatures who have become 'humanised'. Tim Pearson explained:

For a bat to be used for education, and a bat to be in constant captivity and interacting with the public, it needs to go beyond just that: 'Well, if you must poke at my wing while it's sore, or do whatever.' It's got to go beyond that . . . The education programme

[at a nearby wildlife care and tourist facility] means that one of
the daily talks is on flying-foxes, so one of their rangers will get
in the aviary, and then usually bring one of the bats out, one of
the more calm bats out, and show it this far away from people [a
metre or so]. But they also go out in public; they go out to shopping centres on the Central Coast, and schools and stuff. In the
last two years, there's been a Sustainability Fair. I've been there
with flying-foxes, just talking non-stop about how wonderful they
are, and they're threatened, and what you can do to try and save
them. To do that, you need a bat with the right temperament,
because some of them, even though they're comfortable around
humans, will never be comfortable out in crowded places, and so
you need to humanise them, so yeah, it's just basically handling
them, getting them used to noise, getting them used to being put
in a cage and travelling in the car and going out in public. . . .[38]

Face and interface

In encountering an ethics whereby people respond to the calls of
flying-foxes, I am drawn to Levinas's work on the face. He said: 'the
face is the other before death, looking through and exposing death
. . . [T]he face is the other who asks me not to let him die alone, as if
to do so were to become an accomplice in his death. Thus the face
says to me: "you shall not kill".'[39] The related injunction, as I've said,
is that others must not be left to die alone, must not be abandoned.
As Judith Butler reads Levinas, this plea awakens us to the precariousness of the lives of others, and thereby to the precariousness of
all life.[40]

In pressing the significance of the plea not to be abandoned, I am
moved by how Levinas subtly reminds us that actually and ultimately
we cannot prevent the deaths of others. In practical and beautiful
ways, however, we can refuse to abandon them. Sometimes, in fact,
we may even be able to help them return from the deathzone. The
call of those in peril expresses their longing always for connection
within the world of life, and so we are doubly responsible – first, we

have the responsibility to *hear* that call, and, second, we have the responsibility to *respond* to it.

Call and response thus require a double competence: that of calling, and that of apprehending the calls of others, as Dominique Lestel tells us.[41] He concludes that for humans, imagination is a crucial attribute. Especially for those of us whose lives have been shaped within a culture that separates us from nonhuman others, we must first be willing to imagine that others do call, and that whether or not these calls are specifically addressed to us (and most are not), many of them do indeed concern us.

Scholars have queried whether Levinas's concept of the face, the source of the call into ethics, actually includes creatures other than the human. Recent work shows that nonhuman creatures, and even rivers and other elements of the biosphere, can be understood to have a face in Levinasian terms, and thus to call us into ethics. This scholarship importantly emphasises that others call not only in their suffering, but equally in their beauty and their joy.[42] With flying-foxes, the 'face' might include a leathery wing and a gently rasping little tongue, or it might be a million-strong sortie across the evening sky. We are called to exercise our imagination across the differing bodies and worlds and modes of communication, and to recognise our connected lives, connected responsibilities and connected fates. Inevitably, the human condition becomes ethically bound to all these vulnerabilities; in a biosphere of entangled con- nectivities, this is how life lives. Ethics precedes individual life.

Carers in general cherish the unknowable subjectivity of others, and they cherish the fact that at this interface of precarity and ethics, flying-foxes too are active moral participants.[43] We saw this in the cases of mothers adopting orphaned babies, to offer just one example. This is to say that flying-foxes not only receive the care their suffering calls forth from others, but also offer care as it is called forth.

As I came to understand carers' depth of commitment and com- passion, they are deeply certain of the ethics of their work, but they do not, in general, engage in abstract justifications of their commitments of time and energy. They are of course well aware of the public discourse about the need to save flying-foxes from extinction, but on a day-to-day basis the carers I interacted with kept the focus on a few fundamentals: flying-foxes are sentient, they suffer, carers can help. And so they insist on both the uncertainties and the blessings of their work. The call to alleviate suffering is undeniable for them, and their response needs no justification. Indeed, I sensed that for many people an invitation to offer a justification for their response may have seemed superfluous, even obtuse. Their focus was to respond from within an already understood ethics. In Denise's words:

> The thing that really upsets most people is the cruelty of the injuries, and the duration. Somebody's been hanging on a fence or laying on the ground with their wing ripped off and badly burnt. That's what normally does people in . . . I try not to think too much about that . . . Our priority is always the animal. We try to take ourselves out of the equation and do the best for that bat.

Carers rarely talked about the future, although from time to time a person would express dismay and/or anger at the way the odds are stacked against flying-foxes. At the same time, they would not stand by and do nothing. I am impressed with the powerful strength of this ethic of response, and my attention continues to return to the fact that carers insist on treating each flying-fox as an individual subject. These individuals were not objects of management, or pathetic creatures worthy of condescending pity.

Thick care is 'a vital affective state, an ethical obligation and a practical labour'.[44] Among these affective states are modes of interaction that, on the human side, involve love. On the flying-fox side, we cannot know if love is an appropriate word, but their passionate commitment is amply evident and deeply moving. We know that their own modes of communication are at times extremely intense, and some carers have been privileged to become part of that encounter. The experience of being held in an intersubjective cross-species gaze has the potential to offer privileged involvement with flying-foxes' own mode of intersubjectivity.

It is not only that a human addresses the subjectivity of another being, but also that sometimes the other responds. From subject to subject, back and forth, across creatures who give and receive, a truly intersubjective dynamic of encounter arises. Louise was eloquent, and her words arose from her direct experience: they 'give themselves to you . . . They empty their souls into you.'[45]

4   Participation

Paris, 1930s: Lucien Lévy-Bruhl was publishing large tomes of philosophy dedicated to analysing the intellectual lives of tribal peoples.
He did no field work, but rather drew solely on the reports of others,
and his enquiring mind broke through many major boundaries of
exclusion characteristic of his time. He lived from 1857–1939, and
his story is both fraught and fascinating.

In the 1920s and 1930s, it was still entirely acceptable to refer to
Indigenous or tribal people as primitives, and Lévy-Bruhl shared
that terminology. It is hard to read today. What he did not share
was the idea that these others had inferior mental skills.[1] His conclusions were supremely unsettling, and he would probably be far
better-known today had not his own death, followed by the upheav-
als of World War II, intervened.

The Law of Participation

One major strand of Lévy-Bruhl's challenge to the paradigm of
European intellectual superiority was that Indigenous people's
thought 'had its own characteristic organisation, coherence and
rationality'.[2] Unlike many of his contemporaries who worked on
the assumption that Indigenous thought had failed to develop,

Lévy-Bruhl understood that he was encountering a different but coherent logic; it was not oriented towards abstract principles, but rather constituted an attentive matrix gained through experience of participation in the complexities of the actual world of life.

The term Lévy-Bruhl came up with to describe this logic was the Law of Participation.[3] David Abram, a great student of animism, summed it up in these words: 'ostensibly "inanimate" objects like stones or mountains are often thought to be alive ... particular plants, particular animals, particular places and persons and powers may all be felt to participate in one another's existence, influencing each other and being influenced in turn'.[4]

Lévy-Bruhl was iconoclastic on two fronts: the first big provocation was to claim that connection is or can be a form of logic. Secondly, he contended that western people could learn from Indigenous people about this important form of logic. That is to say, the logic of connection is, at the least, complementary to western logic and may in some aspects or contexts be superior. This claim upset the powerful binary that held civilisation (particularly as the property of the West) to be opposed to the primitive (particularly associated with tribal peoples who were being colonised). The structure of hyper-separation conveniently asserted the right of civilisation to dominate.

Central to his analysis was the way the logic of connection upset several key tenets of western thought: including its abhorrence of contradiction and paradox (the either-or logic), its focus on stasis and predictability, and its claims to universality. In contrast, Lévy-Bruhl contended that tribal people's experience of nature is fluid, and that in their actual lives they experience the power of creation as a process of flux and metamorphosis. 'Neither living beings nor objects are monomorphic' and 'the extraordinary is part of what happens normally', he wrote.[5] He thus radically unsettled western ways of knowing and offered a glimpse into an ontological-ecological terrain of mutualisms, flows and co-becomings. Rather than a hard boundary between humans and others, Lévy-Bruhl's account of an alternative logic asserted that living creatures are substantially implicated in each other's lives: they participate in shared life, and their lives are shaped in part through their interactions.

The Law of Participation entails ethical claims, bringing us into domains of connectivity, responsibility, accountability and commitment. Lévy-Bruhl was gesturing towards the logic of 'both-and', the logic of connection rather than exclusion, and of differences organised into multiplicities; both-and welcomes uncertainty, and enables an ontology of becoming rather than being.[6]

Lévy-Bruhl worked out his ideas in his Paris armchair. There was a noticeable lack of participation! Our questions today take us into the heart of lived experience in order to ask: what is the Law of Participation in the actual life, law and culture of contemporary people? What forms of connectivity entail participation, and how are they patterned across time, space and the lives of living beings? What does it mean to experience the power of creation as a participatory process of flux and metamorphosis? To engage with these questions, I draw on my own participatory ethnographic research.

Multicultural worlds

In 1980 I travelled from the USA to Australia to find an Aboriginal settlement where people would allow me to learn about their way of life by living with them and participating in community life. The people who agreed to host my ethnographic research lived in Yarralin and Lingara, two small communities located in the open savannah woodlands of the Northern Territory, and within the watershed of the Wickham River.

Yarralin people established their autonomous community in 1973 after they had walked off the cattle stations (ranches) that had dominated their lives for over seven decades. In 1980 a few members of this group decided to start another community, Lingara, where they could be in closer contact with their own country. The Lingara group (locally known as a 'mob') were particularly welcoming and I learned a great deal from them. I went back and forth between the two communities, as did they, and in my writing I refer to them collectively, for convenience, as Yarralin people. This region was colonised by the cattle industry, and the older Aboriginal people I met had survived massacres, dispossession and ongoing cruelty. Women, men and children had worked for decades without pay. They were given modest amounts of food and clothing, and they were also given a measure of protection from the predatory violence of frontier capitalism, including state-managed punitive control over most aspects of Aboriginal people's lives.[7] Individuals were not citizens, they did not have even the basic protections of citizens, and for decades they were the unfree and unpaid backbone of the cattle industry. As good workers in a skilled and physically demanding industry, they were respected; at the same time, and by the same people, they were denigrated as Blackfellas. They were given the autonomy to pursue their own ways of life seasonally, when they were not needed as a labour force, and so their own country-based knowledge informed their lives. Within people's own lived experience, they necessarily came to understand the injurious contradictions of this enclosed bifurcated social world: as Aboriginal people they were the ultimate others against whom white settlers measured their superiority, and as such they were always refused a place at the table. At the same time, the two groups lived in close proximity, and there were allowable, albeit often contested, types of intimacy, primarily between white men and black women. People in both groups shared a thin layer of culture focused on cattle, and on local knowledge of landscape, climate, placenames and language sufficient for communication. Both groups recognised unspoken rules surrounding their separation and their intimacies. Violence was pervasive but was not the only story, and kindness existed from time to time, along with respect for skill and strength.[8]

The people I met at Yarralin were survivors in that very precise sense of being people whose forebears had been slaughtered. As a mode of resistance as well as a matter of survival, they had maintained their inner dignity and a vast domain of knowledge and remembrance, phrased often as Law. Fidelity to Law was and is the ground of resistance. There was much to resist. Genocide and ecocide have been the paired modes of devastation inflicted on Indigenous people and country in many parts of the world, and because people and country are so deeply intermeshed, harm amplifies into wider ripples of degradation and suffering.[9] Along with attacks on people, there had been attacks on native animals and plants; some were deliberate, and many were side effects of settler methods of land and water use. There was much loss, including local extirpations. Yarralin people continue to oppose the ongoing war against nature. They look at country and see waste on a massive scale. The man whose driving energy made the shift to Lingara possible, Riley Young, gave tough and evocative expression to this experience: 'Why don't you [whitefellas] think back to that Law? You've been enough wasting, shooting people from his country. Because White men come out and made a big mess now. You can see: paddock gone [eroded], grader gone cutting all around. Wasting. Wasting ground.'[10]

Aboriginal people became citizens in 1967, and with greater freedom, they were planning to make better lives for themselves, their young generations and their country. At that time the Australian nation was promoting measures of restorative justice, including land rights, that encouraged optimism. People's first priority once they were free of the oppressive and restrictive conditions that had marked the previous decades was to live in their own country, close to their ancestors, their sacred sites, and their sources of life and meaning. Law and responsibility were located in country, and they needed to be present.

The fidelity that had sustained people through the long years of occupation was focused on the wide group of 'countrymen': the term includes men as well as women, nonhumans as well as humans, the dead as well as the living. When the Lingara group moved back to their own country they were, for the first time in close to 100 years, able to live not as subordinates of whitefellas but as *ngurra mala*, glossed as 'boss for country' or 'traditional owners'. The oldest of the men explained the ethic of fidelity that held them in place and gave meaning to their lives: 'the Wickham River is filled with blood of Blackfellas killed in those days. Their bones are all broken up along the bottom . . . We are camping now on the blood of Aboriginal people killed in those days.'

For the most part, it is the senior people in Aboriginal societies who have the responsibility of teaching others. I was deeply desirous of being taught, and many of the Lawmen and Lawwomen took up the task. The people who took me into their protective care were wise and knowledgeable, and very tough. Having gained a measure of freedom, they were most focused on their own objectives, but at the same time they well understood the power of the written word, and the potential role of outsiders who shared their goals of self-determination. Older people had been denied western education, and while they sought to ensure that their children would learn to read and write, they also enlisted willing outsiders into the project of making their stories, understandings, commitments and aspirations more widely known. The Aboriginal Land Rights (Northern Territory) Act 1976 (Cth), and Aboriginal Sacred Sites Protection Legislation (NT, 1979) opened up new cross-cultural legal encounters in which anthropologists could put their learning to work for Aboriginal people. In the end I did not go back to the USA; there was too much good work to be done in Australia.[11]

Yarralin itself has changed enormously since my first encounters starting in 1980. There are more people, more houses and other amenities, and many more outside officials.[12] Government policies have shifted from the empowering vision of self-determination to a darker model of discipline and punishment. Yarralin is not a happier place, but it is still a place of strong people with strong knowledge, and incredible determination.[13]

* * *

I arrived in Yarralin in September when the sun was intense and the savannah glowed golden and silver. Dark lines of dense growth etched out the billabongs, rivers and creeks. All that brightness! I was captured by red soil and mesas, by glimmering waves of grass, sparkling trees, and by the huge sky that held it all, and that seemingly would go on holding it all, forever.

One of the great delights on those hot evenings in 1980 was watching flying-foxes come forth from their camp further upstream. Their huge bat-like shapes stood out sharply against the deepening sky as thousands and thousands of them flew above us, following the river for a while, and then fanning out over the vast open woodlands. The Eucalypts were flowering, and the flying-foxes were feasting on nectar. I had read about these creatures in various ethnographies, but I hadn't quite imagined what it would be like to live in their neighbourhood. I was awestruck. With time I learned about their place in Yarralin people's cultural world. As Dreamings they were shape-shifters, metamorphic creators, and much of the country around Yarralin and Lingara was marked by their tracks, stories and sacred sites. I came to recognise flying-foxes as totemic kin and as ecological communicators, and to appreciate the unique enthusiasm they bring to life. I joined some of my hosts in eating them from time to time, until in due course, as I will explain, I stopped eating them.

During one of these exquisite flyouts, the Lawman Daly Pulkara pointed out something interesting. The crowd flew over, and then, after the majority were well on their way, a few turned back. A bit later, stragglers appeared, following the others but not quite catching up. Daly explained: 'the old people always said those blokes forgot their axes'. I felt like a bit of a straggler myself as I tried to catch the drift of his words. I had to remind myself that in the beginning, in creation, they swapped back and forth between human and flying-fox forms, and when they were men they would have carried axes. 'They're always like that', Daly said, 'one or two are back behind. That's why the old people said they forgot their axes, they had to go back to camp and get them.'

Sentience is integral. In the literature, Indigenous people's recognition of widespread sentience is often termed animism. The term conveys both an account of life and an ethics for life: animism is defined as the recognition 'that the world is full of persons, only some of whom are human, and that life is always lived in relationship with others'. It follows that the appropriate mode of engagement with nonhumans entails 'learning to act respectfully (carefully and constructively) towards and among other persons'.[14] Within an animist understanding of reality, flying-foxes are persons, and like any group of persons there is variety; some are a bit sloppy, dragging along behind, forgetting their axes when they go out foraging, always a bit out of step.

Yarralin people use the term 'culture' to name these specific ways of life. Mindful creatures have and live in their own manner. The evidence all around them shows that other beings have and follow their own lifeways. They have their own foods, foraging methods, forms of sociality and seasonality; they have their own languages; if we cannot understand their languages, this is not surprising; we cannot understand all human languages. But it goes further than language. One of the Lawmen explained: 'birds got ceremony of their own – brolga, turkey, crow, hawk, white and black cockatoo – all got ceremony, women's side, men's side, everything'. Here, too, evidence abounds. Brolgas (large cranes) are a great example: they have a red patch on their head, and thus look as if painted for ceremony; they perform a dance that is mesmerising in its intensity. Other animals, other lifeways; the presumption is that while we humans never witness more than a fraction of what others do, we know on the basis of what we do encounter that they lead richly interesting and diverse lives. This is not a matter of belief but rather of observation. Life is immensely diverse and fascinating in this multispecies, multicultural social world. Many creatures are held in high regard because of their unique abilities, and humour is there as well. Indeed, laughing about the odd and sometimes goofy things animals did was part of the pleasure of paying attention to the great diversity of living things.

This multispecies, multicultural world is participatory all the way through. Human subjects encounter other creaturely subjects, and to experience life, to actually *be alive*, is to share in and contribute to the liveliness of the world.

Kindreds

One of the founders of the Lingara mob was a fellow named Old Toby. He was close to eighty years old when I met him, and had lived through a brutal history. A short rake of a man, his bowed legs testified to a life on horseback and implied years of hard work: mustering, throwing cattle, branding, breaking in horses, droving. Like many people in the cattle industry, he walked a bit unevenly as the result of various injuries over the years. With his beat-up cowboy hat, western shirt, blue jeans and smart belt, along with an engaging smile and a certain wariness, he was a classic stockman of the outback. At the same time, he was a knowledgeable Elder, a Lawman, and a leader in local Aboriginal life.

Before I could get to know him well, Old Toby died. And when he died, the word went out: no one was to kill flying-foxes. Not in Lingara, not in Yarralin, and not in any of the surrounding communities where people shared Law and ceremony. It was a serious edict affecting all of us, humans and flying-foxes, and it would last until the bereaved relations told us that we could again go hunting. A righteous penalty for breaking this Law was death. Old Toby was a flying-fox man, which is to say that he was a flesh and blood participant in a cross-species kindred (kin group) comprised of other flying-fox people and flying-foxes themselves. The quality of kinship entailed mutual solidarity. As the group suffered, no further suffering could be inflicted until there had been time to recover.

Aboriginal people in Australia have developed one of the world's most elaborate systems of multispecies kinship. Anthropologists apply the term 'totemism' to these kindreds.[15] The term may be misleading in so far as it might seem to imply that the human group is symbolised by, or in some way represented by, its totem. This is not how multispecies kindreds are understood in Australia. The relationship between human and nonhuman members of the group is, in most instances, one of shared substance or co-substantiality, meaning that some of the substance of their bodies is shared through descent from their common ancestors. Something of flying-fox bodily substance is within the bodies of flying-fox people, and vice versa.[16] Their lives are bound up with each other, and what happens to one affects others; risks to one are risks to all, and the well-being of each is enmeshed in the well-being of others. The person who exists in others, and in whom others exist, is vulnerable to what happens outside their own skin, but, equally, they find their power in the relationships that are situated beyond the skin. They participate in the well-being, as well as the suffering, of kin.

An excellent definition of Aboriginal Australian totemism is that it connects beings through 'bonds of mutual life-giving'.[17] Participation in these powerful bonds can be understood through the logic of multispecies participatory flows. A first principle is connectivity: life always depends on and is lived within relationship with others. A second is that the mode of relationship is kinship – there are those to whom you are related, and there are many others to whom you are not related. Third, the encompassing frame of kinship articulates an ethics – there are mutual responsibilities across species and other beings. Fourth, kinship is expressed in structured bonds of enduring intergenerational, interspecies participatory solidarity.[18] Kinship is both a structure that is perpetuated through time and an ethics of practice that gives substance and meaning to the structure. While the structure is founded in descent, the substance is always being formed anew through nurturing care. The process is circular: bonds of mutual life-giving congeal as kinship, and kinship calls forth bonds of mutual life-giving.[19]

The system I discuss here is not the only system in Australia, but it is especially interesting, I think, because of its cross-cutting patterns and connections. Through shifting patterns, the system resists self-enclosure and holds itself open widely and unpredictably to ongoing life. It positively revels in diversities. One type of multispecies kin groups is country based. These kindreds started with the Dreamings and their original work of creation. As the great creation ancestors travelled, they left groups of descendants who were situated in country and are now known as countrymen. For example, from the Emu ancestors come emu humans and emu birds – the emu countrymen; from the Possum ancestors come humans and possums – another group of countrymen, and so on across Australia. The system in which everyone belongs to country and to a group of countrymen has the potential to produce a set of competing singularities with boundaries that obstruct flow and limit participation. However, this potential is not realised because any given boundary is cross-cut by other boundaries. This point may be clearest when we think of the straightforward example of marriage and descent. Countrymen do not marry each other but rather must look to other countries for husbands and wives. Two people who marry are different; they must be different in order to marry. We can think of them as Emu man and Possum woman. As with incest in strictly human terms, marriage requires moving outside the natal family. In generational terms, the children are Emu in relation to their father and his country, and they are Possums in relation to their mother and her country. The outcome is a series of country-based groups: the nightjar people and birds who belong to the country sung into life by the Emu Dreaming ancestor, for example. Equally, though, this unity is not hegemonic. It can be broken up, and other unities formed; in this case, amongst possums.[20]

Alongside the country-based kindreds, there is a different, strictly matrilineal, system; they are differently embodied and their focus is not towards country. The two types add to the complexities of interacting kindreds but do not compete; they mesh at times, but function with different effects. Country-based kinship carries rights and responsibilities that are physically located in country and borders. The matrilineal type is located in bodies; it entails rights and responsibilities located in embodied kinship and broader scales of ecological interactions. The flying-fox kindred, of which Old Toby was a member, is matrilineal. Local terminology for this type (*ngurlu*) indicates 'flesh' or 'body'. The flying-fox members of the kindred are all *warrpa* – Blacks (*P. alecto*). A person is born a flying-fox because their mother was a flying-fox. All of a woman's children share her flesh, but only her daughters pass it on to the next generation. Here, too, people must find spouses in other kindreds. This system is focused on the actual fleshy bodies of the individuals who instantiate the kindred. As a general rule, people do not eat their nonhuman countrymen. But nor do they hoard. Rather, human responsibilities for the well-being of others include the obligation to perform ceremonies and other actions that keep the nonhuman members of the kindred healthy and thriving; the work benefits the kindred, and it benefits other humans in other kindreds. Everyone has to eat, and hunting is a joy, but at the same time everyone knows that the creatures they eat are someone's kin. The obligation is thus to refrain from being wasteful or disrespectful. In a world of kin, care of one sort or another is always required; others are paying attention. Furthermore, if humans believe that their kin are not faring well they have the right and duty to prohibit hunting until the species' prospects improve. Thus, in addition to the relatively random prohibitions brought about by human death, there is also the work of monitoring nonhumans and managing humans.

* * *

The animal who is your kin is called your *warpiri*. In the context of *ngurlu*, Daly Pulkara explained the meaning this way: *warpiri* is your 'biggest sorry' (your greatest sorrow). In my words, *warpiri* is YOUR GRIEF. Your flesh and blood, flowing through the bodies of others who will be hunted, is *your* grief. You don't have to grieve over everything; you couldn't. But here, in this place of encounter with the necessary deaths of others, your grief may become extreme and so may your anger. If these deaths come about in defiance of Law, righteous anger is directed to those have not managed death properly.

The question thus arises as to what it means to manage death properly. As we saw with Old Toby and the flying-foxes, a hunting taboo is placed on the *ngurlu* species after the death of a human member of the kindred. This was serious, and the Law was respected. I was involved with one event that focused my attention on breaches of Law. It happened shortly after the death of a prominent emu man in a nearby community. The recently deceased emu guy was not old, and he had died suddenly; people were still getting over the shock. Throughout the region no one was to kill emus until the emu people said it would be okay. Deep in their grief, they had not yet lifted the taboo. But, on a trip from Lingara to Yarralin we spotted an emu near the road and pulled up because one of the men, a flying-fox guy named Morgan, wanted to take a shot at it. The Elders were against this, but Morgan grabbed the gun and shot. The emu was wounded, and the children all jumped out of the truck and raced to kill the animal. After considerable excitement with sticks, stones, guns, a wounded emu, shouting children and barking dogs, the emu was 'finished' and brought back to the truck. All the way home Morgan protested that he had not meant to shoot the emu. The Elders were worried, and their discussions focused on chastising Morgan, and on figuring out whom to give the emu to. Its death had been unlawful; this was not a random act of violence; emu people's grief would soon turn to anger and all of us in the truck were to some degree implicated. Morgan was way outside the Law; we in the truck were involved by proximity (at the least), and we were in difficulty. Embroiled in Law-defying action, we were wrong-footed, culpable and exposed to angry grieving people who wanted vengeance. We couldn't hide the dead bird, nor, worse yet, throw it away. We needed help from the very people whose kin had been killed because they held the power of life and death; their anger was wild in its intensity and would quickly become focused on its target (us). In the end we took the now dangerous emu body to one of the oldest Yarralin Lawmen, who fortunately had an emu connection through his wife, and he undertook the perilous work of butchering and cooking it.

Morgan always maintained that the death had been an accident, and when others suggested that it was not much of an accident to have loaded the gun, pointed it at the emu, and pulled the trigger, he reminded people that the gun had no sights and that its value in hunting was a complete joke. It was a pretty steamy time while people considered Morgan's actions. He was respected in most other contexts of his life, and over time people stepped back from holding his impulsive actions against him. But in that time when the Law was breached and the anger was palpable, reparative action seemed to be a far horizon. His own kin had to acknowledge that he was in the wrong, and at the same time they had to find ways to extricate him from life-threatening peril. The emu people had the right to kill him with impunity, as he had killed one of their kin. The negotiations towards reparations and reconciliation were complex and were not part of the public record.[21]

Clearly, kin groups may become sites of violent emotions and actions, and, inevitably, life is never free of human passions including those that are harmful. Perfection is not a goal and nor is the suppression of human emotion and vitality. The key point is that, broadly and sustainably, the effect of these great concentrations of diversities and responsibilities is to spread care across many species in a structured manner from which, in the best of times, everyone benefits.

* * *

The system of matrilineal kinship, *ngurlu*, is not specifically country based, but rather opens out to wider cosmic scales and relationships. By 'cosmic' I refer to events, processes, actions and relationships that unfold widely across earth systems and beyond. The movement of stars and constellations, for example, and the pulses of drought and flood: there are actions and patterns that while experienced locally arise from much wider systems. One of the main wide-scale divisions is between rain and sun, wet and dry. A person on the rain side would preferentially marry a person on the dry side, thus bringing together the two great climatic forces that generate the flourishing health of country. Flying-foxes are a great example: they are intimately connected with dark rain and thus with the rainy season, and are complementary to dry season kindreds such as, for example, emus. Ultimately, they are deeply associated with the powerful, world-sustaining Rainbow Snake, the great driver of rains and the generation of life. As I will discuss shortly, they are implicated in the motion of seasons, the flowering of savannah trees, and the health and nutritional generosity of country. Thus, through a multiplicity of forms, and multiplicity of scales, life comes forth in dizzyingly diverse participatory proximities of cross-cutting, kin-based, co-substantive bonds of mutual life-giving. Participation is complex and is always oriented both towards others as well as towards one's own. In this system, living beings truly stand or fall together.

* * *

Inevitably, I came into Yarralin and Lingara as a stranger. This is not a category that people felt at all comfortable with, and almost immediately they gave me a classificatory identity that would position me so that we could actually interact respectfully. Deeper relationships took longer, but while I was in that first loosely constrained identity I was widely free to act, learn and explore. I could eat whatever I chose, for example, and I enjoyed learning to cook and eat foods I had never encountered, including flying-foxes. The vast majority in our region were Blacks, and although the Little Reds visited occasionally and were well known through stories, the focus of food, kinship, country and interaction was all with the Blacks. People hunted them with shotguns or batted them out of the low trees along the riverside when they were camping there. Dead flying-foxes were first singed to remove the fur, and then were cooked slowly on the ashes. I never developed an appetite for the brains and guts that some of the old people relished, but the meat was tasty, and the whole process of sitting on the ground around a pile of ashes, cooking, eating, chatting and sharing was a delight.

All that changed when the senior woman at Lingara, Mir Mir, claimed me as her sister. She was the wife of the storyteller Daly Pulkara, but more importantly she was the flying-fox matriarch. As her sister I too became a flying-fox woman. Our children were all flying-foxes, and kinship, including our connections with rain, seasonality and the Rainbow Snake, added an experiential awareness of, and interest in, the comings and goings of our nonhuman kin. Now, for me, these beautiful creatures were in fact food no longer. They were spectacular flyers, and the subject of interesting stories, but more than that they were family. The stragglers who forget their axes were not just oddballs, they were our oddballs, the silly ones that show up in every kindred. Kinship became more than classifications and rules; I began to experience and enact an ethics for life in a multispecies world.

Law and Dreamings

The Yarralin Lawman Hobbles Danaiyarri offered an explanation of origins which shows the matrix of creation, continuity and Law: 'Everything comes up out of the ground – language, people, emu, kangaroo, grass. That's Law.' There's nothing in this statement about rules, sanctions, norms, obligations, punishments or property. In fact, only one element in the statement directly concerns humans. We are alerted to a very close connection between Law and country.

Law is central to Aboriginal culture. It came into being by Dreamings, and it continues. It concerns the life of the world: how things fit together and how they continue. To come up from the ground is to be part of creation: Law and creation are intertwined. And because creation was not only about the emergence of humans, but about the emergence of everything, Law is intertwined with everything. And because creation is the emergence of patterns and connection, tracks and sites, kin groups and rules, Law forms the basis of connectivity and relationality. Law gives the conditions for mutually beneficial participation.

Dreamings established countries, speaking the language as they travelled, and leaving humans and others to take care of it. A country is small enough to accommodate face-to-face groups of people, and large enough to sustain their lives; it is politically autonomous in respect of other, structurally equivalent, countries, and at the same time is interdependent with other countries. Each country is itself the focus and source of Law and life. One's country is a 'nourishing terrain', a place that gives and receives life.[22]

The creators were shape-shifters, changing back and forth between human shape and the shape of the creature they also become. They sang their way across the country, imprinting it with form and character, making sites and songlines, organising kin, country and ecological connectivities. Not only were Dreamings shape-shifting between animal and human forms, but other elements of earth life were ancestral as well. The Sun and Moon Dreamings were part of creation, and sites for both of them dotted the Yarralin area. So, too, were numerous beings that western ontology does not recognise. The Rainbow Snake was one such figure, powerful and active in creation, powerful and active still today.

Dreamings travelled, and from time to time they stopped. When they stopped, they acted in ways that left physical remains, or they transformed themselves into permanent sites, and they stayed. These places are now known in English as sacred sites. Equally, they kept going. Dreamings are the ultimate nomads, masters of an art that includes both motion and stasis, departure and return. They are both here and there, fixed and mobile. And they are both then and now: both origins and contemporary presence. People interact with Dreamings in daily life as they do their hunting, fishing, gathering and visiting.[23] And people interact with them especially powerfully in ceremonial contexts. The Law people – men and women, human and nonhuman – continue to sing the songs and participate in the ceremonies. In the everyday life of birth (hatching, spawning, sprouting, and so on), as well as in the highly charged participatory encounters achieved in ceremony, creation is the always-coming-forth. Creation flows into our lives today because continuities are sustained by everyone's participation in Law.

* * *

A brief detour into ceremony as I experienced it at Yarralin helps understand Dreaming action. Ceremony gives us glimpses, and perhaps this is exactly how our understandings of participatory creative action should be: forever incomplete, but always vivid when present. For myself, I learned experientially that understanding arises episodically. A flash of insight arises in one moment or place, and it can connect with other flashes, forming patterns that may become generative into the future. At Yarralin and surrounding communities where I have danced on many occasions, one of the main ceremonies is called Bandimi. The songs sing up the track of a group of Dreaming women. The men sing and the women dance. I learned to dance, and so I learned to work the ground with my feet and learned to make the dance-call that is integral to the pattern. Thus, I learned that the body connects Earth and air when you dance. The call comes from deep within and is propelled by the impact of your feet on the ground. It comes to feel as if the ground itself propels your voice out into the night sky. That call starts somewhere below your feet and ends somewhere out in the world. The call is a motion, a sound, a wave of connection. You are dancing the earth, and it feels that the earth is dancing you, and so perhaps you are motion, a sound, a wave of connection.

The ethnomusicologist Cath Ellis uses the term 'iridescence' in describing Aboriginal music, with specific reference to the Pitjantjatjara musical system. She explains this unexpected concept with reference to the phenomenon that occurs when background and foreground suddenly flip. Everyone experiences this phenomenon in visual form, particularly with art or photos that are designed to generate the flip between background and foreground. The flip phenomenon is also experienced aurally, as one or another pattern is heard as foreground, similar perhaps to choral singing of contrapuntal music, where cantoris and decani exchange foreground and background. The song becomes iridescent through the complexities of the shifting ground of interweaving patterns.[24] In Bandimi we women danced all night while the men sang. Each segment of song and dance, however, was set apart by a counterpoint of non-dance. Each small song was punctuated by a break. The rhythms of the song and dance were thus set within a larger oscillation of music and non-music. The non-music interval was dedicated to joking. It was not a break in the ceremony but rather a contrapuntal engagement with the musical portion of the ceremony.

When I danced I also experienced my own embodied iridescence. There is the flip between the feet on the ground and the ground on the feet: who is the dancer and who is the danced? It is all in motion, and the significance of this mutuality is located in the flip back and forth between us.[25] In ceremony one becomes part of the pattern, and to become part of the pattern is to join in the call. In that place of iridescence, a further question arises: who is calling and who is called? To become part of the call is also (when things go properly) to become part of the response. One flips from calling to being called, back and forth all night long. To dance, therefore, is to move within a generative, liminal matrix of betweens – between the ground and the foot, the earth and the air, between the many inter-locking patterns and flips, and between the enduring Dreamings and our ephemeral lives.

Iridescence is a mode in which one performs patterns and takes them apart again, performing both departure and return. As we withdraw and make separate, so too we return and interpenetrate. We embody and perform life's desire, reiterating it again and again, and in doing so honouring and participating in the ripples and waves of life's gleaming, motion-filled coming forth.

Tracks, sites, stories

It is not possible to offer extensive information about the Dreaming Flying-foxes in the Yarralin region because so much of their action is bound up in secret ceremony. Male Flying-foxes are associated with circumcision ceremonies as well as with higher orders of men's initiation and with major regional ceremony. The Dreaming men were warriors and Lawmen. The women were Lawwomen. Each group carried their gendered secrets that constituted knowledge originating in ancestral Dreaming creation. Secrecy is a relational term: who is on the inside and who is outside of a boundary relat-ing to knowledge that is not freely available to all. The gender-based boundary at the core of secrecy enables women and men to acknowledge each other as holders and managers of essential knowledge without which life could not go on in the mode of ini-tial creation. Other factors are relevant: age, the tracks and sites that define country boundaries, and individual progression through stages of knowledge and ceremony participation. My focus here is on the specifics of non-secret information concerning Flying-fox Dreaming action in the Yarralin region.

The Flying-fox Dreamings came from the north-west, walking along the high stony country at the top of the watersheds. The men carried their axes and spears, the women carried coolamons and baskets. They were talking and singing, and the men were ready to fight. At various sites along their route they stopped for ritual. Their track went parallel to, but mostly stayed clear of, another prominent Dreaming in the region: the Black-headed Python. She, too, travelled from the west, but where she carved out rivers and creeks, the Flying-foxes kept their distance, staying up in the hills.

At an earlier stage in the Python's travels, there were actually two sisters, but one of them fell ill. When that happened, they parted company. One of them kept travelling east, and the sick one turned off to the south. Near the place where the sisters split up, the male Flying-foxes split up too, one group going east with the great Python. She is a founding figure in a major regional ceremony, and Flying-foxes were and are key participants. The group which turned off to follow the sick Python found, when they got to her, that she was dead. They wept, then, and did the 'sorry business' for her there at the top of a creek that is part of the Lingara mob's country and is cared for by them. The Flying-foxes' choice to stay clear of Pythons makes sense as pythons are one of their main predators. According to scientific research, 'most permanent flying-fox camps have their resident python. These snakes climb trees and appear to mesmerise the flying foxes. The pythons grab the flying foxes and quickly crush the body which is then swallowed whole, head first.'[26] It seems all the more interesting, then, that a group of Dreaming Flying-foxes carried out the mortuary rituals for the dead Python, crying over her body and singing her back to her origins. Participation, it is clear, is not confined to interactions that are comfortable or that hold no challenge. More significantly, relationships that may be antagonist in everyday life are brought into complementarity in the context of ceremony. Thus, the male Flying-foxes and the female Python are both key Dreamings in a regional context.

The Flying-foxes who continued to travel got to a high, flat-topped mesa near the junction of the Wickham and Victoria Rivers where they stopped and marked a boundary. From here the men threw spears at other Dreamings in the area, telling them to stay back, to mind their own business and their own boundaries. Having established their own limits, they turned towards the place where Yarralin is now. The men went up a small pocket to a pointy hill to set up a men's camp and start doing men's ceremony. The women went on to the billabong near Yarralin and made a camp for women and children. There they rested, bathed and ate. Some of the women went off to a nearby place where they did women's ceremony.[27] Later, the men went back up into the stony country to join the group that was grieving for the dead Python.[28]

There is a sacred site which is of such significance that no human beings are allowed to visit. In fact, people are not even allowed to drink the water in the area without first being ceremonially intro-duced by a senior traditional owner. The site is a big tree located on the bank of the river and is described as 'the mother of all the flying-foxes'. I speculate that this sacred site may be an enormous maternity camp, but of course I do not know. The human response to declare off-limits a place that may be a maternity site is, at the very least, full of respect for the gendered work of flying-foxes. And yet, accounts of flying-fox origins are curious, and never specify the mammalian mode of giving birth.[29] Yarralin people explained to me that flying-foxes spend part of the year underwater with the Rainbow Snake and indicated that they may come into being as offspring of the great serpent.

A written account may seem detached and dependent on a bird's-eye view, but the experience of travelling through storied country, where the evidence of creation is all around one, and where one is at home, offers an emplaced immediacy that connects past and present through country, Law, the actions of living beings and the actions of Dreaming ancestors. When we drove to Yarralin and Lingara we travelled the gorge and looked at the boundary hill and the pointy hill. Arriving at Yarralin, we knew that the billabong was just a short walk away, and that the Flying-fox women had stopped and rested just over there. We lived with and within the evidence of all this action.

* * *

Another connecting link between past creation and current life is that Dreaming action set out templates of form and behaviour that continue into the present. The flying-fox stragglers who forget their axes are doing what some of their Dreaming ancestors did. As it was, so it continues. The story announces the behaviour, and both story and behaviour continue from these origins.

Other stories of how things come to be as they are concern antagonisms between megabats and microbats. All of these creatures are represented as stroppy, argumentative, quick to offer insult and quick to take offence. Most of all, though, nobody seemed to like the Little Reds because of their different smell. Inevitably, fights arose: Blacks fought with microbats, for example, and their 'soldiers killed a big mob' of the little guys. One story has the Blacks slanging off at the Little Reds, accusing them of lurking around with sexual promiscuity on their minds, sneaking up on other women, and having a great old time with them. The Blacks smelled them and reckoned they'd kill them for that. They accused them of unlawful sex, and of course the Little Reds didn't like to have that said. The Blacks 'made them ashamed' of their own smell. Interestingly, in stories told by Yarralin and Lingara people, the Blacks always won.

Then there was the conflict between a Little Red and a microbat. Once again, the Little Red was being abused because of his smell. The Little Red grabbed his spear and went after the microbat, and when the little guy tried to hide amongst some rocks, the Red poked the spear up his bum. That microbat still has the shape of a creature with a spear in its backside, and it flies in an erratic path as it seeks to evade the angry Red. This microbat is an important

Dreaming figure in its own right; it is described as a small bat that lives in hollow logs, and it is perhaps the northern freetail bat or a sheathtail, both of which camp in hollows and have a protrusion that resembles a spear in their backside.[30]

Excursion into vulnerability

The metamorphic flows which are sustained through participation are life enhancing, and at the same time perilous. When damage occurs, it expands. I learned about this through an encounter with terrible harm. The Northern Territory Aboriginal Sacred Sites Act came into effect in 1979, and Aboriginal people quickly began documenting sacred sites and nominating them for registration. I assisted in this process, working with groups of countrymen to visit sites and prepare the formal documentation required to support the nominations.[31]

We got a huge shock in 1986 when we drove past an old homestead near Lingara and saw that all the big, white-barked eucalyptus trees near the homestead had been chainsawed into oblivion. They were Dreaming trees where a fish Dreaming known as Jajiki (probably the banded grunter) had transformed. We had documented the site for registration in 1982, and we were stunned to see the majestic trees wantonly destroyed. They lay on the ground, their leaves still shining with the particular dusky glow of their kind. Around us the grasses were still yellow, the sky still immense and blue, and other trees still stood in their places. Along the horizon the mesas stood immobile, as they always do, their steep sides shimmering with the silver haze of the smoke trees.

We stood there for a long time taking it in. The senior Lawman for these trees and this country, Daly Pulkara, had a stunned and anguished look on his face – the kind of look that tells you that a person has just had the guts kicked out of them. I got the video camera and asked him to speak about the destruction; I was already imagining us in court prosecuting the people who had committed the crime. My shameful eagerness to pull the event into a context in which I thought we would regain some agency was horribly disrespectful to Daly and to the trees, and, as it turned out, was also pointless. We were told later that there had been a bureaucratic error and that the station owners had never been notified that the trees were registered sacred sites. They may not have known to hold back on the chainsaw, and so could not be held accountable for their action. Over the years, we had documented numerous sites in this area, and Daly had told terrible stories of participatory suffering: how Aboriginal workers had been ordered to chop down some Dreaming trees, for example, and how the trees screamed when the axe went into them, and how the men who did the chopping died off one by one in quick succession. I suppose all this was in his mind as he looked at the new disaster, gazing at the country that had witnessed all this. 'It hurts people's feelings', he said, generalising the experience. Then he claimed it as his own and spoke directly to the perpetrators: 'You hurt me', he said, 'you take my power away'.

Feelings, in Daly's Aboriginal English, are experience. Hurt feelings constitute the feeling of being hurt, of being attacked, cut, chainsawed, hacked at; the feelings include loss and grief, the pain of knowing it shouldn't have happened this way and that nothing can undo the damage. I knew he was seeing the deaths of cherished others, trees and people, and later I thought that perhaps he was seeing his own death too. Daly knew the story already; he had seen and known such interrelated deaths, and these trees were his kin and his responsibility. He had a stroke not long after this event and was never again able to speak well. Often, we just sat together looking at country and watching the kids play. He died a few years later. Before the stroke, though, Daly amplified his sense of devastation and ongoing loss:

We'll run out of history, because Whitefellas fuck the Law up, and they're knocking all the power out of this country.[32]

Communication

The Wickham is one of the great untrammelled rivers of North Australia, and magnificent trees thrive along its banks. River red gums (*E. camaldulensis*) and paperbarks (mainly *Melaleuca leucadendron*) along with figs and Leichhardts tower above the river and provide shade all year round. There are smaller pandanus palms, freshwater mangroves and viney undergrowth. Out in the flat country away from the Wickham and its tributaries the savannah woodlands ripple with grasses and are dotted with trees, predominantly large Eucalypts.[33] Further away from the river there are dry and stony mesas with their own spinifex grasses and smaller Eucalypts that thrive in such conditions. The country is classed as semi-arid, a category that doesn't fully acknowledge the great diversity from one year to the next, some of which are exceedingly dry, others exceedingly wet.

The monsoon tropics are marked by two main seasons – wet and dry; in addition, there are two transitional phases, the build-up to the wet, and, later, a cooling-off period before the dry. When I first encountered this startlingly intense country in the late dry season (September) the Wickham was not flowing. For months it had been without replenishing water, and the dry riverbed was punctuated with a series of shrinking waterholes. Each of these sites of permanent water had a name in local geography; each was an oasis of shade and life-sustaining water during that part of the year when the country was becoming ever-more hot and desiccated, dusty and faded. As the dry season moved towards the build-up we entered a period of increasing heat and humidity. Summer was coming to the southern hemisphere and the monsoon was building in the Indian Ocean. Temperatures rose up into the 40s (105°F+), and it felt as though the country couldn't get any hotter or more stifling, except that it did. The ground was too hot to walk on comfortably in bare feet, all the smaller waterholes had dried out, and the atmospheric pressure was dense. It seemed to go on like this forever, until finally the rains came and started to cool the heated ground.

The rains were sporadic at first, but soon the river started to flow again. Its first flush was terribly hot and dark as the water came running off the dry ground. New growth started up almost instantly, fresh green grass sent up shoots that quickly became tall grasslands, and over the next few months the country was revitalised. The billabongs and small ephemeral waters were refilled, and the river flowed mightily. In an occasional super-wet year it left its bed entirely and came out across the flats. With rain the white-barked Eucalypts gleamed from being washed, and the open country started to sparkle. And then the rain stopped: grasses turned yellow very quickly, and the water remaining from the wet season would have to carry the country through the long dry and the hot build-up until there would again come rain.

The great seasonal forces are for Yarralin people expressions of the power of ongoing creation; they are part of the ontological-ecological terrain that works with patterns, communication, connectivities and the ongoing flows of life. Wet season and dry season, the great life-shaping powers, wrestle back and forth: Rain and Sun, Sun and Rain. In the logic of both-and, they are both antagonistic and complementary. Summer coincides with the wet, and winter coincides with the dry: the wet is a hot season, but is cooled by rain; the dry is a cool season, but is warmed by sun. Living beings have learned to live with extremes, from the desiccated aridity of the late dry to the swampy ground and rushing rivers of the wet. You could die of thirst, or you could drown, each possibility is totally real and almost every year in North Australia a few people do actually die in these ways.

In creation stories the Rainbow Snake is the great being associated with all water. I refer to it as 'it' because its gender is too complicated to be reduced to either 'he' or 'she'. Throughout Australia the Rainbow Snake (or Serpent) is recognised as the great and powerful figure associated with water, and frequently, also, with male initiation. Descriptions vary from place to place. A painting of the Rainbow Snake on the wall of a rock shelter near Yarralin shows a large, large snake with characteristic protuberances on the head,

Figure 4.1 Tracing of Rainbow Snake, Yarralin region
Source: Traced by Darrell Lewis, from rock art near Yarralin.

often described as ears.[34] It is described as 'boss' for rain, and in the Yarralin region it is associated with every permanent spring and waterhole. Where there is permanent water, there is the Rainbow Serpent – in the rivers, in the aquifers, and in the action of the rain itself that brings the fresh water down to Earth: dangerous, powerful, sentient and enduring. The fact of permanence in this country of change is living proof that something powerful is there. Furthermore, the Rainbow Snake embodies the idea that water is *both* a powerful presence *and* an ethical subject. In calling it an ethical subject I mean that the Rainbow Snake is enmeshed in, and responsive to, calls for care and responsibility, including care for flying-foxes.

Across much of North Australia flying-foxes are intimately associated with water and the Rainbow Snake.[35] Yarralin people used the word 'mates' to describe the relationship. After gaining an understanding of heat stress and why it is that flying-foxes are so dependent on water, I could discern an ethological element to this partnership. No water, no flying-foxes. For Aboriginal people in the monsoon tropics the dramatic action of water, as rain, as cyclone, as rivers, creeks and billabongs, its coming and going, demonstrates its centrality to all: no water, no life.

The seasonal pulse of blossoms is matched by a seasonal pulse of flying-foxes, and the pulses are accompanied by, or announced by, communicative events. Towards the end of the late dry season, the balance had shifted towards intense heat and the country had become almost unbearably hot. After months of foraging in the

Figure 4.2 Flying-fox figures in rock art, Ngarinman country
Source: Photograph by Darrell Lewis.

open country, the flying-foxes came to camp near permanent water. They hung from the low-growing pandanus trees, primarily, and they supplemented their diet by munching on leaves of a riverside shrub that Daly showed me; the little flying-fox tooth marks were clear.[36] A flying-fox camp is invariably lively; there is always lots of pushing, prodding and scrabbling about. In these riverside camps it was always possible that an individual might lose its grip and become a tasty treat for crocodiles. For their part, the crocs waited just below flying-fox-laden trees for an unlucky individual to tumble.[37] It was a risky camping place for small edible mammals – one false move and you're gone. Why did they do this? One reason, I now understand, is that they needed the water and humidity to counter the increasing heat. The reason the Yarralin people offered was that they came to address the Rainbow Snake, telling it to get to work. They called out to their mate, urging it to get up and get going, to bring on the rain. Others joined in: the frogs shouted their crazy chorus, waterbirds came flocking in, cicadas were shrieking. This intense communication became very noisy, and people too added their voices. If all went well, the Rainbow listened and rose up; it towered over the earth, emitting lightning, thunder and rain. The first rains started to create moisture that moved back into the sky, forming more clouds and more rain.

There is a multispecies ecological pattern here. My pre-eminent teacher of botany, Jessie Wirrpa, divided the Eucalypts into those that flower in the dry time and those that flower in the rain time. The savannah trees which flower in the dry time include the spectacular inland bloodwood (*Corymbia terminalis*) and the magnificent tree known in vernacular English as the half bark (*C. confertiflora*). These species produce large, showy and heavily scented flowers and are thus obvious attractors, along with several other species including *E. pruinosa* (smoke tree), which grows out along the sides of stony hills, and *E. microtheca* (coolabah) which grows around billabongs. As Jessie explained, they flower in succession from higher ground to lower ground, from the drier country on the hillsides down to the riverbanks and channels. River red and the paperbarks are outstanding examples of rain time trees. They burst into flower in one final extravagant outpouring of vitality at the end of the sequence. Flying-foxes follow, and here at the riverside, Yarralin people said, they talked to their mate the Rainbow, telling it to bring on the rain. The succession from dry to wet was complete. The blossoms were finished, and there would be no more until the rains renewed the country.

This is a story of communication: how trees call to flying-foxes, announcing their nectar, and how flying-foxes call for rain. More seductively, it is a story of desire: how flying-foxes and trees want to live, how they attract and benefit each other, and how they participate in each other's lives and renew themselves. It is a story of ethics: how the call for life renewal is answered with rain. It is thus a story of mutual flows across species and through time. It may be a bit of a stretch to imagine that flying-foxes and Rainbow Snakes share a language in the way human speakers can share a language, but by this account, to act is to communicate, and communication has, or should have, consequences. The Rainbow Snake is responsive; it experiences the calls of others, and those calls awaken it to its responsibilities in the world. And so, across this continent of heat, dust, drought and fires, and through all the El Niño climate warping fluctuations, the rains do come.

Participatory solidarity

On a fairly recent visit to Lingara, Mir Mir was away and her eldest daughter Aileen was the senior flying-fox woman in camp. After I had talked with everybody, Aileen took me to the bank of the river for a chat. She often did this: holding my hand and taking me away from camp, into country, side by side looking at the river, to share news of what was happening. This time her gentle voice carried a lot of anger as she talked about the latest thing whitefellas had done. They'd come through Aboriginal communities, including Lingara, telling everybody not to touch flying-foxes: not to eat them, not to handle them or hunt them, not to have any physical contact at all because flying-foxes might have disease. I could understand something of what was troubling her. Terror, it should not be forgotten, is exercised as a multispecies project. Aileen's grandfather had been shot and killed in the early days, some of her relations had been poisoned, many members of her family had seen their dogs brutally shot, and dingo poisoning was a continuing feature of the war against animals in this region. Her father, Daly Pulkara, had had a stroke not long after the Dreaming trees had been chainsawed. All of them – people, animals, plants and landforms – had felt the oppressive weight of life and death decisions made elsewhere and imposed ruthlessly. This new public hygiene advice looked like another round of colonising violence. That's what her tone of voice seemed to imply.

Aileen was not going along with it. 'They've been here forever', she said of her flying-fox kin, 'just like us. We're not worried. They're family.'

* * *

The Law of Participation, as Lévy-Bruhl termed it, opens into a world of multispecies connectivity, kinship, ethics, responsibilities and mutualisms. This is not soft and fluffy; there is a certain harshness here. Vulnerability, accountability, knowledge and care are not up for grabs. And so the disrespect for Law which pushes some people, like Morgan, to or beyond a Law edge has social and ethical consequences. It may become extremely difficult to sustain a life worth living. At the same time, participation, understood as interactions and flows of mutual benefits, including care, brings the meaning of life into everyday action, articulating relevance between Dreaming creation and the day-to-day lives of many species, including, but not dominated by, humans.

To return to the metaphor of a wave, creation grounds this ontological-ecological terrain, and participation brings living beings into actions that ripple across daily life, continuing creation and at the same time renewing possibilities for all that may yet come forth.

5   Nomads

Flying-foxes are nomads. Unlike migratory animals, they do not travel together in one large group, and nor do they follow regular migration routes. Rather, their travels are individualised and episodic; their nomadism is both regional and long distance. When there are large amounts of food they gather in great numbers, and when the food is dispersed they too disperse. Their travels are thus perfectly attuned to the episodic and unpredictable qualities of Australia's boom-and-bust ecologies. Camp sociality is spatially and temporally fluid. Some flying-foxes are always leaving. Some return and others arrive from elsewhere, bringing reports of what they have encountered. Information exchange is significant and is facilitated by the diversity of individuals' travels.[1] Their stories are both localised and detailed, as well as extensively pathed. Sometimes, with tail winds, they travel faster than they can actually fly under normal conditions (25–30 kms per hour).[2] As an earthbound mammal I can only imagine how thrilling it must be to race over the country in fine flight.

Travel

Humans have always known that flying-foxes are great travellers, but only recently, with the invention of lightweight satellite telemetry, has it been possible to track and measure the movements of individuals over very large distances. The evidence is consistent: perhaps not all flying-foxes are great travellers (we don't know), but those who undertake long-distance journeys are indeed adventurous spirits. A point of caution: only individuals can be monitored with a device; we do not know if these individuals travelled alone, or if they were with a group. If they were with a group, did group membership shift along the way? One more caveat: it may be that the experience of being fitted out with a tracking collar is sufficiently unpleasant to convince an individual that it is time to get a move on. We just don't know. If, as has been suggested, there is a survival advantage to long-distance travel, then it is reasonable to imagine that most flying-foxes are wide-ranging.

On the basis of existing evidence, three patterns of travel stand out. The first is the pattern of the seasons and is linked to the phenology of preferred plants. In general, large camps are formed over spring, summer and autumn and are then broken up during the winter when smaller numbers of flying-foxes move around from food source to food source. These relatively local travels are well documented through radio telemetry. The second pattern is that of the flying-fox life cycle. In the spring, large maternity camps are formed. After three months or so, the babies start to fly, and although they may be nursed by their mothers for a bit longer, they are becoming independent youngsters. At about this time, the mating season starts. The second part of the life cycle pattern, then, is the mating camp where the next generation is conceived.

The third pattern concerns what we might fairly call wanderlust. It involves long-distance travel according to a pattern that hugs the well-watered coastal region. Journeys cover distances of 1000+ kms or more, one way, along the coast, while ranging approximately 128 kms between coast and inland.[3] Two scientists, Tidemann and Nelson, undertook a satellite tracking study of two Greys, one in a camp in Melbourne, Victoria, and one in a camp near Tweed Heads in the far north of New South Wales. They found that each of these individuals travelled at least 1,000 kms from their camp of origin, and each returned to their camp of origin. It should be noted that while these two individuals went travelling, not everyone in their camp of origin took off, so it is clear that the camp had not become uninhabitable.

The first guy was fitted out with a tracking device while he was in the Melbourne Botanic Garden. He was monitored as he travelled along the coastline, first going east through Victoria, and then heading north. He camped as he travelled, of course, and he came all the way to Sydney. While in the big city he visited several urban camps, including the magnificent Gordon camp, and later he turned course to go home, ending up back in the Botanic Garden. The round trip was well over 2,000 kms.

The second individual was fitted up for tracking at a camp in far north NSW. He stopped short of Sydney, but he also made an interesting, perhaps unusual, excursion up into the mountains. He, too, made a round trip of over 2,000 kms, and returned to the place where he was initially collared. The camp sites of the second guy were correlated with camps in that region, and it was found that many of his stopovers were at sites that were occupied at the time. He was visiting, in short. Many of the populous camps were widely spaced, so other stopovers were necessary, and he may have 'spent the day alone or with a small group'.[4]

Tidemann and Nelson propose four potential reasons for long-distance mobility: (1) additional food; (2) mating opportunities; (3) information about other parts of the country; and (4) a combination of all three.[5] The fourth point is clearly the most persuasive, and if we were to translate these points into everyday language we would suggest that adventurous flying-foxes sally forth from the security of the home camp in search of new foods, more opportunities for sex, and the thrill of exploring new country. To this we might further consider the factor of information sharing. The individual who travels gains knowledge. It would almost certainly enhance an individual's personal reputation, which may be linked to their social/sexual desirability, to be known as someone who has a lot to communicate.

The evidence from this study has been borne out subsequently, and the accumulating data indicates that flying-foxes just love to travel. Ongoing research shows that it is not only males who go off on great expeditions. A recent study of a female Grey followed her for an amazing four years, and the map of her travels is astounding. It shows her criss-crossing the narrow band of country along the east coast in dense clusters of activity. During this time she travelled between Gladstone (Queensland) and Bateman's Bay (NSW), a distance of about 2,000 kms, but as the available map conflates data it is not possible to develop an idea of what any given season or year would have been like for her.[6] Her story turned toxic when she arrived in Bateman's Bay in May 2016 for a great flowering of spotted gums. She was not alone. Reports from communities in the area suggest that perhaps 100,000 individuals came to feast, and that many of the human residents were irate. The local shire was reported to have received $2.5 million to torment the creatures and thus, it was hoped, to disperse them.[7] Subsequent reports say that the numbers have dropped enormously. Was this an outcome of dispersal, or was it life as usual for flying-foxes who moved on to other food sources once this one was finished? Most probably an answer would include both outcomes, and others as well.[8] The adventurous female who was collared for such a long time appears to have disappeared from the story.

Long-distance travel is undoubtedly captivating, but it must not be thought that the flying-foxes who stay in one place are therefore strictly sedentary. They fly out at night for food, and they visit. A radio-tracking study in Sydney led the scientists to conclude that 'colonies are dynamic rather than static assemblages' and that 'adjacent colony sites may effectively behave as one site, with part or all their populations shifting' back and forth 'for purely local reasons'.[9]

Navigation

The most astounding thing, for me, in these stories of long-distance travel, is not so much how far individuals went, but the fact that they returned to the place where they were first collared. Their navigational skills are clearly superb. Local manifestations of navigation are well documented. Hall and Richards tell us that youngsters are taught navigational skills by older flying-foxes, and there are many observations of nightly flyouts in which great mobs of creatures follow rivers, or streetlights and other urban landmarks, as they go about their nightly foraging.[10] Olfactory navigation also plays a role. One of my many thrilling flying-fox moments was when I was coming home on the train just at flyout time. The train was well above street level at that particular place and a flying-fox flew alongside the train as we pulled out of the station. This urban individual cruised just above the mess of electricity wires, following the street running parallel to the tracks. I gazed with delight, while he or she was focused on flight; all too soon the train went underground and the moment was over. Sydney's transportation system did most of my navigational work for me, while the flying-fox negotiated a complicated geography of wires, lights, elevation and lethal obstacles. To me, the work appeared effortless. Fleeting, unpredictable and suddenly present, such glimpses of multispecies proximities and possibilities have the potential to grab a person and to open into adjacent worlds of meaning and action, but one has to be willing to respond.

Navigation is widely construed as the process by which 'a course or path from one place to another is identified and maintained'. It is 'based on the capacity to plan and execute a goal-directed path'.[11] Ants and honeybees, wasps, birds and all mammals, including humans, along with many more creatures: we all navigate. We go out for food, and many of us return home bearing both food and information. As mammals, flying-foxes have the capacity to navigate using a cognitive map. Birds also have this capacity, and there is debate about whether insects, for example, may too. Much depends on how one defines the cognitive map; my perspective is that it is more insightful to hold to a narrow definition rather than trying to include all creatures who move about with purpose. The differences are real.

In a recent study of wayfinding, John Huth analyses three main elements in human and other mammalian cognitive maps: dead reckoning, perception and proprioception. They build on each other.[12] *Dead reckoning* (or *path integration* – PI), is a term many of us humans associate with sailors. It entails a process whereby knowledge is built up by accumulating 'successive small increments of movement onto a continually updated representation of direction and distance from the starting point'. Travelling into the unknown, creatures build up knowledge of that journey. This type of navigational skill is fundamental to animal navigation, from 'molluscs and insects to humans'.[13] *Perception*, as Huth uses the term, involves the ability to perceive the surrounding environment, and thus to navigate using landmarks and geometric relations; another term is piloting. The creature who pilots sets a course towards a goal on the basis of landmarks that are already known, and it monitors those landmarks as it travels. That is, having mapped a new route, the creature remembers and follows it. In contrast, *proprioception* adds to perception; it involves skills that depend on the experience of the body. A creature experiences their own sense of motion and distance; you feel yourself running, for example, and know that you are covering more distance than if you were walking. Proprioception, therefore, adds embodied experience to the map, enabling a creature to build up an overall understanding of terrain.

Huth makes the important distinction between route knowledge and survey knowledge. *Survey knowledge* is a key feature of cognitive maps. This type of knowledge depends on a creature's capacity to understand a terrain as if it were being seen from above and integrated into a whole that is greater than the sum of known parts. It includes both route knowledge and the broader spatial knowledge that enables an understanding of a given terrain.[14] Survey knowledge is thus integral to how creatures figure out how to make short cuts, navigating not only by landmarks but by a knowledge of how known elements fit together. This mode of knowledge implies a grid that includes both the known and unknown, and it enables a creature to race off into the unknown and incorporate new knowledge into wider and more extensive maps. This capacity to expand from the known to the unknown means that a cognitive map can be developed to include unfamiliar geography. New knowledge can be tested in flyouts; it will not necessarily prove accurate, but it is integral to a cognitive map. In contrast, *route knowledge* depends on known paths and landmarks. Having learned specific routes, a creature stays with them. Extrapolation to the unknown is not part of this mode of mapping. Creatures such as ants depend on route knowledge whilst lacking survey knowledge and therefore do not possess cognitive maps.[15]

It is now possible to locate these functions within the brain. Huth tells us they dwell within the hippocampus, an old brain area shared amongst mammals.[16] Navigational functions are activated into two types of cells. Place cells fire when a creature (such as a lab rat) is in a particular place, and they fire again when the creature enters another place. They are integral to both route knowledge and survey knowledge. Grid cells actually make a grid of equilateral triangles and enable creatures to formulate survey maps and thus to locate themselves spatially in ways that go beyond what place cells do. Huth also points to declarative memory which he defines as combining these navigational skills with the ability to recount memories verbally.[17]

Flying-foxes are mammals like us, and they are great navigators, relying on both route knowledge and survey knowledge. We know that they communicate navigational knowledge amongst themselves, and so we conclude that the ability to share spatial memories includes non-verbal ways of knowledge sharing.

Excursion into wayfinding

I experienced the significance of the cognitive map when I lost mine. The lesion on my skull was pressing into my brain, and over the course of a few days I completely lost my spatial grid, along with numerous other functions. For a short while I could barely even access route knowledge. Crucially, I also lost the capacity to analyse the fact that I'd lost my grid. I can reflect on it after the fact, though, and think about how I lost my navigational skills, and how, as the pressure was removed, I came back into mammalian life.

I have experienced the feeling of being lost in the bush, and it is nothing like losing one's cognitive map. Having the hippocampus under pressure and unable to do its navigational work left me bereft. It became hard to shop, for example. I followed others, getting a shopping trolley and watching where they entered the store. The sliding doors didn't look like anything I could recognise until I saw people going through them. Similarly with the check-out area: I knew what I needed to do, but I had to follow other people to get to the right place. A few days later I went to my partner's university to attend the ceremony at which he was awarded his doctorate. I was not familiar with the campus, and fortunately a mutual friend helped me. While she thought she was helping me get oriented, I was mutely aware that nothing she said made sense. What I had to do was memorise a couple of routes using landmarks that had strong visual distinction. To get from the coffee shop to the ladies' room was my big challenge. The first time my friend took me, the second time I had to do it on my own. From where I sat with my coffee I could see where the ladies' room was, but the only way I could get there was the way my friend had taken me. So: step out in the right direction, leave the wicker chairs behind, wend my way through the odd-looking metal tables and seats, work around to the ramp that took me up to the right level, and look for the symbol for ladies. Coming back, I reversed this, feeling grateful that the key features were quite distinctive.

By the time I got to hospital the next night I couldn't even follow landmarks. The physiotherapists had me up and walking within a day or two. The aim was to get my coordination working again. I didn't tell them how desperately confused I was. We walked a circuit that took me to the family room and then back to my own room without turning and retracing my steps. I tried to memorise landmarks, but they all looked the same, and the only thing I could really hold in my mind was that my room was next to a big mural with yellow flowers or leaves.

Over the course of a few days my navigational capacity returned. I no longer needed landmarks because I understood how it all fit together. I found that I had been following an unnecessarily long route to the family room; that was how I was first taken when the point had been exercise, not navigation. With the return of my cognitive map I could choose my route. This little corner of the ward was, I should say, not much larger than a three-car garage.

Interestingly, when I was in 'ant mode' the distances seemed great. My path was a series of landmarks, but I didn't have an internal sense of how far apart they were, or indeed how many there were. It was simply a matter of putting one foot in front of the other, one landmark at a time until I got to the place I thought I was aiming for. In retrospect, it seems that my sense of distance and speed was connected to my spatial grid, and without it, proprioception was disrupted. I knew I was going slowly because that was all I was capable of, but I couldn't relate that fact to distance or time, and thus to a sense of how long it took to get somewhere.

Over the course of a few days I became a mammal again, a human being; indeed, I became, once again, myself. As I reflected on the experience later, I thought: I've gone from being a mammal to being an ant, and I've come back again. There was quite a bit of truth in this thought, but it was unfair to ants because they have a keen 'homing' capacity. They go out, and they may try a new route, but they always know how to get home again.[18] I couldn't even get to my hospital room without concerted, stressful effort. To paraphrase Karl Marx, what distinguished the best of ants from the most impaired of humans, based on my experience, is that they are perfectly suited to their way of life, having all the intelligence and navigational skills they need to be the complex creatures they are. I was suited to nothing, sustained only by medical science and my own determination to improve. The experience of de-evolving and then returning to my own kind of world brings me into ever more gratitude for medicine, and the mammalian brain, and the resilience of the human body.

Site fidelity

'We tell ourselves stories in order to live', Joan Didion famously tells us. As humans we understandably become focused on our kind of stories, and we may forget that other creatures are also telling themselves stories in order to live.[19] The flying-foxes which travelled a thousand or more kilometres in order to visit new country ultimately turned around and headed back to the exact place where they had started. They did this through their navigational abilities, and with a narrative of home place that engaged their own life histories and desires.

The technical term for returning home is site fidelity, and in much of the flying-fox literature it is also termed roost fidelity.[20] Natal site fidelity is the most common type: animals return to their birthplace in order to breed. Site fidelity involves a nomadic story of departures and returns. Love of travel is counterbalanced by love of home place. To leave requires a story of venturing out in response to information such as flowers, or perhaps being gripped by a desire to go elsewhere. To return requires a story of home place and why it matters, as well as how to get there. Stories and cognitive maps – each separately, and both together – depend on time-binding: the mental process that connects one event to another, linking them into a sequence with meaning. Time is part of 'the subjectivity of life, the first person experiential world'.[21] Time-binding enables creatures to become enmeshed within a flow of time, to be in a 'now' that has a past and a future, to make meaning and to become part of an intersubjective world of meaning-making. It is to join an always-unfolding story. Among social animals, stories are both individual and shared. Flying-fox travellers who have been tagged indicate through the course of their lives that their stories were undoubtedly personally compelling, and at the same time were part of wider, shared stories about birth, place and the significance of coming back.

Paul Shepard offers an avenue into understanding the way in which many nonhuman animals render their experiences and perceptions as successive and meaningful events. His example concerns predator-prey relationships. A predator necessarily constructs a narrative when it is tracking another animal: where did the animal go, how fast, what kind of condition was it in – all bits of information (among many more) that can be developed into a story about the prey animal and that can lead to strategies for a successful kill. Similarly, a prey animal also needs to be able to construct a story: does the available information indicate that a predator is nearby? Is it moving towards or away from the prey? On the basis of the available information the prey has to make life or death decisions: shall it hide, run or play dead? Both predator and prey need good cognitive maps: where is the terrain dangerous, where would a creature hide or take a short cut, what evasive actions, or surprise actions, are possible given the existing possibilities?[22]

The predator-prey analysis clearly shows how cognitive maps and narrative fit together in the lives of many animals, including humans. Flying-foxes, while not exempt from becoming prey, are active at night and thus protected from their mostly diurnal predators. Humans are exceptional predators, as we manage to be active 24/7 if we choose. More significantly, flying-foxes' focus on blossoms means that while they need their maps and stories the purpose is not to outwit the trees but rather to be fully responsive to them. The relationships are co-evolved and mutually beneficial, as I will discuss shortly.

Hall and Richards explain site fidelity this way: 'One of the distinctive features of flying-fox ecology is their fidelity to established camp sites. Camp sites where young are born become especially significant to those animals, and they will continue to return to the camp, possibly for the rest of their lives.'[23] Some sites have existed for over 100 years, and some may go back to the time before European colonisation.[24] Counts of individuals at two sites in NSW show the great variation between the crowded population of a camp when the young are just starting to fly and its later state after a great number have left for the winter. One site varied from 40–45,000 at the peak to 500–1,500 at the minimum. Another site, one which had been documented to have been in existence since 1852, peaked at 150,000 (approximately) and later was totally deserted.[25]

Maternity camps are chosen by females, and year after year they return to the same place to give birth. We have evidence of their determination to remain in place because of all the human efforts to get flying-foxes to move on. But setting aside human intervention, we also have dramatic evidence thanks to research carried out before and after Cyclone Larry. This category 4 tropical cyclone hit North Queensland in March 2006. The scientists who assessed its effects had been monitoring the population in that area prior to the cyclone. They found that for six months after the cyclone, about 250,000 individuals, or 90 per cent of the Spectacled flying-foxes in the area (*P. conspicillatus*, known in the vernacular as Spekkies) had disappeared. And yet, by November, the numbers had bounced back to about 209,000. Most of the returnees came back to known maternity camps, and these camps became larger than they had been before the cyclone.[26]

A maternity camp, like any long-stay camp, must have water nearby, and there must be a diversity of trees and shrubs with the foraging area sufficient to continue flowering during the life of the camp. There may also be a serendipitous, co-evolved advantage. During the period of time encompassing late pregnancy and lactation, females benefit from higher levels of calcium in their food. Fascinatingly, research in the tropical north suggests that the big riverside paperbarks (*M. leucadendra*, for example) provide high calcium 'rewards'.[27] Reading about this possibility makes me think of the flying-foxes at Yarralin and how they congregated at the riverside; I wondered if there was a hidden bonus. Were pregnant females getting extra calcium just when they needed it?

The females in a maternity camp remain in close proximity to each other and don't allow the males to come too close. When all goes well, a baby clings to its mother as soon as it is born and remains with her full time for several weeks. After that they wait for her while she goes out foraging. After the age of three months the young ones start to separate from their mothers and become

Figure 5.1 Flying-fox mother and young
Source: © Nick Edards.

part of the teenage cohort: noisy, curious, clumsy and full of energy. Mothers continue to groom their young; the separation is not instantaneous, but with time the youngsters become adults, ready to take off on their own adventures in life.[28] Many of them, both male and female, will return to this camp.

There comes a time in January or so when the males who are old enough to mate (about two years old) start to become far more active. Females have mostly weaned their young and a camp that had been relatively quiet while the mothers were birthing and nursing their babies takes on a new aural ambience. Marjorie Beck was one of the key figures in Sydney's Ku-ring-gai Bat Conservation Society; she lived near the Gordon camp, so she was well aware of what this time of year meant in terms of noise. 'The juveniles don't stay out long', she said, 'just a few hours, and when they come home they party all night as they practise their take-off and landing skills and fly from tree to tree gaining strength for the long flights they will need to make in the near future to forage for food'.[29] At around the same time, the older males are establishing their personal mating sites on sections of particular branches which they scent mark and defend against takeovers by other blokes. Competition is extreme. Individual males cut short their foraging time in order to return to defend their site, and they utter loud booming calls that announce their determination to defend their territory and that may also be intended to attract females. The night camp becomes very noisy, and many people find it pretty trying.

If people who are sympathetic towards flying-foxes find noisy camps to be a nuisance, people who have no such sympathy are more likely to start calling for expulsion. For their part, flying-foxes are intensely determined to remain in places of established fidelity, and it may be that this same attachment to place also drives some of the determination they show when people try to get them to move on from any site whatsoever. This is to say that while site fidelity calls individuals to return to their natal camp, or their personal mating places, flying-foxes also experience strong obstinacy about staying at any camp where the conditions, such as food and water, are favourable. The extreme flexibility in relation to travel seems to be matched by an extreme determination to stay in a chosen place. Site fidelity is understood to be a major factor in clashes between humans and flying-foxes.[30] Tim Pearson, ever fiercely passionate, explained:

> I find it very hard to not get worked up and get going but . . .
> Every attempt at relocation has been a failure. The bats have gone somewhere close and have come back. They've got incredible roost fidelity. They come back to their roosts. Basically, the only way to get rid of them is to destroy where they were roosting. . . .
> I mean, short term, you can drive bats out of a colony. Right? There's no doubt about it. You use enough noise, you can drive enough bats out of a colony. As soon as you stop making the noise, they'll come back. They'll definitely come back the following spring, in the cycle. Recently, up in Eton, Queensland, near McKay – there was a problem with bats, so they got a permit to trim the trees they were roosting in. 'Trimming' meant cutting these magnificent trees off to two metres above the ground, so they were just stumps. Hallelujah! The bats have gone! A week later there's an article in the press saying they're back, they're in the same street, they're 150 yards from where they were before, and they're in trees in people's backyards and people are pissed.[31]

This is the dilemma: the very qualities that make flying-foxes the creatures they are – site fidelity for part of the year, and blossom chasing for another part of the year – put them in harm's way if they venture too close to humans. Recurrent togetherness allows social bonds to be strengthened and allows a large range of information to be passed around.[32] And yet, togetherness also has drawbacks. As humans continue to expand their presence, taking up more and more land, flying-foxes are put in ever greater peril. The Little Reds' differing seasonal cycle means that a town or community may experience two waves of visitors if there is a maternity camp in the area.

The extraordinary intimacy of camp life (by human standards) has numerous benefits, one of which may be the experience of security.

Unless disaster strikes, babies are blissfully protected, close to their mother, warmed by her body, nourished by her milk, and wrapped in her wings while she sleeps, protected from rain and from other creatures. At first their faces are always close; their eyes and voices connect intimately with each other, and with no one else. Later they separate, but the first experience of life is deeply protective intimacy within one place. Louise Saunders witnessed the destruction of a camp, attacked with disorienting noise and lights in an effort to drive away the flying-foxes. There were still babies there who couldn't fly well enough to get away, and there were mothers who were so confused and upset that they fled. Of the babies, she said: 'We had a few babies left behind. They were struggling. They didn't know how to fly away. They just knew that this was home and they had no idea where to go.' Mothers who had become separated from their babies returned hoping to find them: 'And you could see them coming and doing circles and circles, working out what they were going to do.'[33]

Human violence forces us to think of flying-foxes as victims, as indeed they are. And yet, when I turn my attention away from cruelty, I am gripped by the vivid intensity of flying-fox life. Imagine, for a moment, the stories they might tell. Geography is part of it, but perhaps there are tales of paradise, where blossoms are bowed down with lashings of nectar; tales of mountains, and cold, and the struggle to cross over into a land of intoxicating flowers; tales of the best sex ever or of partying all night. There will be stories of flying faster than one ever thought possible, watching rivers, coastlines and city lights zoom past below. Increasingly, too, there will be stories of skyscrapers, brick walls and all manner of strange encounters. On the basis of flying-fox actions, they seem to have very little to say to warn each other of danger. But however stories are told they engender the most incredible determination to return to a camp, and, once there, to stay.

Flying-foxes' actions demonstrate such concentrated commitments that we can with justice liken them to love. Think: mums in love with their single-birth babies, sharing a communication register that belongs just to the two of them, and think of how they gaze into each other's eyes. Think: creatures in love with nectar, and in love with sex. Remember: love of travel and the complementary love of home place. Think: pulses of intensity flowing out into the rhythms of departure and return.

Arts of return

My encounters with the flying-fox ethos of travel and site fidelity are undoubtedly influenced by my understanding of human nomadic patterns and ethos as I came to understand them from Yarralin and Lingara people. It is true that colonisation sought (and continues to seek) to force people to settle rather than travel in an effort to make them dependent and controllable.[34] Aboriginal people had never been subjected to this kind of hierarchy, and colonisation was a massive blow to the autonomy people cherished so deeply. And yet, in most parts of Australia people never ceded territorial sovereignty. They resisted the colonisation of the mind, and continued to understand themselves, their country and kin, and their responsibilities in life and in death, in the way their parents, grandparents and ancestors had done, holding fast to that valiant 'forever'. A strand of contemporary identity politics resists the idea that Aboriginal people were nomadic.[35] It appears to be based on extreme definitions of nomadism; one argument, for example, is that Aboriginal people were not 'true' nomads because they didn't wander aimlessly. By cutting out the interplay between travel and home places, such a viewpoint misrepresents the knowledge and planning that are required of humans if they are to live successfully, and to live with pleasure and purpose. I wonder, in fact, given our cognitive maps, whether it really would be possible to wander aimlessly. The place and grid cells would be firing; people would need to take note of the landscape in order to hunt and to locate vegetable foods, and it seems unreasonable to imagine that they wouldn't be cognisant of their own experience and put it to good use. The image of people wandering without attachment ignores the profoundly attentive care that Aboriginal people direct towards their own country, and to their relationships with other countries and other species. It ignores the evidence of countless land claims which focus equally on sites and travels within a context of ownership of country (often glossed as belonging). And thus it pays startlingly little attention to the great achievements of a nomadic ethos, including the maps known as songlines, and their related ecological ethics and connectivities.

In a curious turnaround, many contemporary non-Indigenous people seem to want to claim a nomadic lifestyle, valuing their mobility and their capacity to land on their feet wherever they go. Another strand of contemporary fascination with nomadism is the deep and serious philosophical interest that goes beyond post-modernism to embrace immanence and to resist any sort of hierarchy.[36] That a nomadic ethos exists as a desideratum among non-Indigenous people at the same time that it is rejected by some (certainly not all) Indigenous people, is one of the many conundrums of contemporary life.

I embrace the term nomadism, and I offer some very specific accounts of how people live in known, named, resource-laden country through a culture that is ecologically attentive to, and respectful of, what is happening *in country*. This is so important: an Aboriginal nomadic life is always lived *somewhere*, and as we have seen, and will continue to see, country in Australia is subject to relatively unpredictable pulses and flows that circle through large earth systems and local interactions with the effects of those systems. One of the Yarralin people's givens is that living beings are participants in webs of such pulses. Relationships are knowledgeable and depend on purposeful mobility involving both departure and return. Departure does not lead to abandonment, or to giving up on a place. To depart is, in effect, to promise to return. Conversely, to return is to promise to depart and thus to refrain from using a place too much or refusing to give it a chance to experience recuperative quietude (often the wet season) between pulses of abundance. Both matter. This nomadic ethics works with seasons and with local geographies. To depart is to live lightly, to depend on resources, and then to leave them and allow them to regenerate. To return is to fulfil an implicit promise: that no departure is forever, and every return is an occasion for joy, bearing in mind that a joyful return is only possible if the country has not been trashed.[37] The expectation that there must be country and resources to which one can return, sets a nomadic ethos within time as well as place, linking the care of past generations with the present and with work that seeks to ensure the lives and livelihoods of those yet to come. The aim, for the most part, is to engage in action from which everyone benefits.

One of the best examples of such action is the use of 'cultural fires'. There is excellent evidence to show that Aboriginal burning was integral to the long-term health of Australian flora, fauna and ecosystems.[38] Detailed studies of cultural fires are explicit in showing that such fires were authorised by Dreaming ancestors, were practised year after year by people who truly knew what they were doing at a fine-grained scale, and who intended that the fires would benefit others as well as themselves. For example, a good fire makes for good hunting, and it is not only humans who hunt. Similarly, a good fire makes for excellent regeneration, and new green growth is more nutritious for herbivores than old, dried-out growth. Aboriginal Elder April Bright, whose homeland in the floodplains benefits from regular annual burning, explained: 'If we don't burn our country, we aren't taking care of our country.'[39]

The big picture is that human beings participate in ecosystems in ways that promote mutual benefit across many species.[40] Thompson, a scientist whose research focuses on mutualism and related adaptations, defines co-evolution as a process of reciprocal evolutionary change between interacting species.[41] In more abstract terms, this nomadic ethos of mutual benefit can be termed 'symbiotic mutualism'. Mutualism is a form of symbiosis involving 'interspecific interactions that increase the fitness of both species'.[42] Symbiosis is defined as organisms of different species living together in physical contact.[43] A growing body of research is showing that mutualism complements competition and is utterly fundamental to life on Earth. Thompson's monumental study of co-evolution reminds us that mutualists are part of how every creature lives. He wrote: 'we have learned that mutualisms are not only common but also much more fundamental to the organization of life than we had previously imagined'.[44] In Lynn Margulis's colourful language: 'Not only are our guts and eyelashes festooned with bacterial and animal symbionts, but if you look at your backyard or community park, symbionts are not obvious but they are omnipresent.'[45] No organism exists in isolation, ever. Research into mutualism does not claim that interspecies interactions are consciously attuned to meet the needs of others. Rather, mutualism is adaptive; it is relevant at multiple scales from individuals to groups to species and beyond, and it entails the fact that over the long term, mutualists gain benefits from the interactions.

The mutualism between flying-foxes and their food sources is almost certainly the result of co-evolution, an evolutionary process that began with the ancestors of Pteropids and the ancestors of contemporary Myrtaceae species. As we encounter the relationships today, we see co-dependence that is mutual, co-evolved and beneficial to all. Hall and Richards offer fourteen points of evidence that go to show co-evolved mutual co-dependence. Their points range from the biology of flying-foxes, such as their adaptive tongue shapes, to the biology of myrtaceous plants, such as, for example, those that put their nectar out in the middle of the night, thereby specifically nourishing flying-foxes and other blossom bats.[46]

It is easy to slip into an overly harmonious vision of mutualism, but ecological relationships really hold no place for wishful thinking. Mutualisms are far more significant *as they are* rather than as we might wish them to be.[47] The great philosopher Isabelle Stengers makes it clear that symbiosis is not about 'ideal peace'. It does not involve a hierarchy of commitments, nor does it involve abstractions. Stengers is interested in finding ways to talk about ecological processes that include humans as well as nonhumans. So, in her words:

The 'symbiotic agreement' is an event, the production of new, immanent modes of existence, and not the recognition of a more powerful interest before which divergent particular interests would have to bow down. Nor is it the consequence of a harmonization that would transcend the egoism of those interests. It is part of what I would refer to as an imminent process of 'reciprocal capture.'[48]

The symbiotic agreement, then, requires difference without hierarchy. The difference between flying-foxes and trees is just such a difference. 'Each of the beings . . . [that is party to] this relationship . . . has an interest, if it is to continue its existence, in seeing the other maintain its existence.' It is a process of encounter and transformation, not absorption, in which different ways of creaturely life find shared interests in each other's well-being.[49]

Exactly the same point can be made in respect of Aboriginal nomadism. In bringing Aboriginal nomadism into dialogue with flying-fox nomadism I am working with difference without hierarchy, while at the same time indicating some shared patterns. This analysis risks being misunderstood in two main dimensions: (1) it risks 'primitivising' Aborigines, making them seem less than human; and (2) it risks humanising flying-foxes, making them seem less than what they really are in the integrity of their own life worlds. My intention is to honour both kinds of living beings, and, also, to honour their many mutualists and, in the case of Aboriginal people, their multispecies kin. Let us bear in mind that with mutualisms, no one dominates: no one diminishes others nor is diminished. Aboriginal humans and flying-foxes are commensalists – they share food, and thus at the very minimum they share an interest in the future flourishing of that food. To think with humans and flying-foxes as they engage with a shared food is to open up a study of complex matrices of immanent becomings.

Sugarleaf – multispecies pulses

In life, connectivities go all over the place. It is difficult when talking about complex matrices of connectivity to know where to begin. The temptation is to start with the familiar – with humans, for example, or with flying-foxes. I prefer to defamiliarise the terrain, and so I will start with the stuff that comes out of the tail end of an insect called a psyllid (*Hemiptera psyllidae*). In Australian English the stuff is called by an Aboriginal word: lerp. The Aboriginal English of North Australia names it sugarleaf.

The story, briefly told, starts with the curious effects on Australian trees of the combination of nutrient-poor soils and exuberant sunshine. Because of all the sun, trees photosynthesise far more carbohydrates than they need given that their nutrients are coming from such poor soil. They fill their sap, leaves and nectar with sugar. All these sugars attract a great range of mammals, birds, parasitic plants and insects. Psyllids feed on leaves.[50] They need amino acids, and, like the tree they live on, to get what they need they also get large amounts of sugar. Psyllids exude the sugar and use it to build little shelters for themselves where they are protected from sun and wind. These shelters are food for others. A great many creatures come to slurp the lerp – birds, possums, gliders, flying-foxes and humans, along with many others.[51] Australia is one of only two places on Earth where exudates are important food for mammals. From time to time there are great explosions of lerp that offer fantastic 'feasting and breeding opportunities' for birds and many others. Animals 'get fat' on all the abundance,[52] and there is a report from an early English settler noting that Aboriginal people also 'got fat' when the lerp was abundant.[53]

I was not able to participate in harvesting sugarleaf, so my account derives from what women and men told me. Sugarleaf came forth in the late dry season, and from time to time, certainly not annually, there were super-abundances. At these times people's sugarleaf gathering really came into its own. Older Yarralin people remembered sugarleaf with great pleasure, but the process of gathering and packaging sugarleaf was labour-intensive, and it was women's work. While women remembered their days of getting and processing sugarleaf with fondness, they were not keen to take it up again. 'Too much humbug' was Jessie Wirrpa's down-to-earth opinion. Yarralin people distinguished two types of sugarleaf: one, *jalyingarna*, comes from the savannah bloodwood trees, and the other, *purungun*, comes from river red gums. Both were said to be extremely sweet, but the bloodwood type was said by some people to be a bit 'cheeky' compared to the other. Both could be eaten as porridge, and both could be processed into cakes.

The method went like this: using a stick with a special hook, women broke off branches of the sugarleaf-laden tree and left them on a cloth (or on large pieces of paperbark) to dry out. Then the branches and twigs were removed. Threshing was done by striking the remainder with the stick to loosen the lerp from the leaves. The remainder was winnowed by being tossed in the air. Once the sugarleaf was separated and cleaned up for eating it could be cooked as porridge for immediate consumption. In a year of abundance, vast quantities of surplus sugarleaf were harvested, and the method of storage was to use small amounts of water to form cakes similar to damper. The cakes were allowed to dry out, and were then wrapped in cloth or paperbark. Sugarleaf cakes could be kept for months, perhaps much longer, and so they formed an important part of the stock of food staples that carried people through lean times. One of the women who knew it well said: 'Oh, yeah, you can keep it months and months, that sugarleaf. In a basket or anything. You can keep it a long time. And you can take it a long way.' One man who had lived in the bush away from white people for many years, and who was well into his eighties when I spoke with him in the 1980s, explained to me: 'that's what I lived on, that sort of tucker grew me up. No bread, we lived on that kind [of tucker].' Another man spoke lovingly of it because it saved his life as a baby. His mother died when he was very young and the other women fed him on sugarleaf, using their own breast milk to make a porridge suitable for a young child.

In years of great abundance, sugarleaf provided a huge storable surplus, with the result that the people who had harvested the abundance could send bags of cakes to other groups who participated in the regional trade networks. Alternatively, they could save it for when they hosted a ceremony, inviting others to join them. Regional gatherings were dependent on there being food for far more people than the local area could normally feed. Sugarleaf was one of several abundances that enabled large-scale ceremonial life, replaced today by the large number of loaves of bread that women bake using flour and yeast. Flying-foxes were a source of the meat that was needed to feed large gatherings. In more recent times, a bullock does the job.

Sugarleaf is associated with women more strongly than any other food.[54] The stick women used for breaking off branches was a dangerous tool, brought into the region by Dreaming women. Men were not allowed to touch it. Moreover, up in the rough hill county not far from Lingara there is an area known as women's country. It is a large area, sacred to women and prohibited to men. A man who fails to observe the restrictions, whether accidentally or deliberately, is in grave danger; this area is women's country, absolutely. It is said to be rich in sugarleaf, but men are not allowed to eat the sugarleaf that is gathered there, at least not under ordinary circumstances. In thinking about the strong relationship between women and sugarleaf, I wonder if a year of sugarleaf abundance, with its enormous injection of calories into people's diets, pushed up women's critical fat levels, resulting in a greater than usual number of pregnancies.[55] This hypothesis, which is speculative but based on relevant evidence, would mean that pulses of sugarleaf abundance led to pulses of human abundance.

Let us shift focus and look at this matrix of mutualities from the perspective of a bloodwood tree. Bloodwoods are open savannah trees; they flower in the late dry season, as I have said, and they are seasonal signals, indicating that hot weather is soon to come. Yarralin people use the wood of this tree for spears, and the ashes are good to mix with chewing tobacco. Some trees develop a hollow that can hold water, and so a bloodwood could save your life. Bloodwoods are one of the main hosts to orchids, and orchids are a source of poison that can be used to put down a rogue dog. These magnificent trees are also home to native bees that make nests that people raid for honey (sugarbag) and wax. Furthermore, the trees produce a red sap that is medicinal. Bloodwoods and other trees host mistletoes, and these hemi-parasitic plants provide homes and food for birds, possums and many others.

Many of the creatures whose lives are enriched by bloodwoods also enrich the life of trees. The animals who visit or live in the mistletoe let loose a downfall of nutrients so that each tree becomes a powerhouse of nutrient recycling. Flying-foxes pollinate, and so, by day, do birds and insects. Humans don't do a lot for bloodwoods these days, but formerly the fires people lit to clean up the country had the effect of keeping the savannah woodlands open.[56] With the suppression of cultural fires, the introduction of cattle and the whitefellas' use of bloodwood trunks as fence posts, the ecology of the savannahs is changing, and with climate change and related factors the future is truly unknown.[57]

Formerly, patterns of abundance rippled across bodies in space and time, bringing forth new lives in waves of co-generative interactions. Every encounter, still today, is embodied history: each mutual touch, each thrust of tongue into blossom with nectar received and pollen deposited, is a moment of mutualism that emerges from the past, captures and anticipates multiple becomings, not only of flying-foxes and trees, but of all their consociates, including human beings.

* * *

The nomad ethos I came to understand in North Australia is attentive and responsive to patterns of departure and return. Pulses add other rhythms of fire and flood, for example, and of wet and dry seasons, of outpourings and of quietude, shaping the experience of life with patterns of departure and return. And indeed, almost everything in country is departing and returning: animals, berries, nuts, grasses, sugarleaf, flying-foxes, winds, and much more. This is multispecies co-generative work in action: ripples of abundance are flowing across species and across organic and inorganic systems, as participants call and respond, arrive and depart, nourish and are nourished. All this flow carries and goes on waves of multispecies, intergenerational mutualities.

And so, with all these calls and responses, a nomadic ethos brings us into a realm of ethics. Here one cannot speak for all creatures; their responsibilities are their own and some may indeed work within the realm of ethics or constitute a 'proto-ethic',[58] but if we hold the focus on humans, we encounter a fascinating terrain. To respond is to affirm the connectivities that make ethics possible. As James Hatley tells us (see Chapter 1), creation brings forth life that is always indebted, always called to and for, always active and interactive, always offering potential for mutual benefits and care, and always with opportunities to say Yes! – and thus, in the mutuality of the well-being of others, always potentially open to the ethics of responsibilities to others and to life more widely.

6 Ancestral Power

Exuberance

She was cruising. Her new baby was tucked beneath her wing, holding onto her fur and her nipple, as she followed the enticing scents of blossoms. In thinking about her life, I draw on facts we know about flying-fox ethos and intersubjective behaviour in order to embark on an empathetic thought experiment that can take us more deeply into the story. I imagine her little one is a boy, and we know that he will be unceasingly attended to by his mother for a few weeks. During this time, he is learning lessons of co-becoming that are both elementary and profound. Indeed, it is astonishing to think of the social and cultural intensity he experiences in the course of just a few weeks of total dependence. Who are we? How do we live? Who am I?

They fly out together, my imaginary friend and her young son, and he learns fundamental life lessons whilst nestled beneath her sheltering wings. He learns about flight itself – about the embodied experience of airborne motion. He senses the thrill of travel, and the fact that there's a destination, and thus a purpose, to all this joyful flight. Mother and baby leave the home camp in company with others, perhaps thousands of others. So the little one experiences departure en masse: he becomes part of a stream of others also launching into flight with destinations on their minds. The baby's first encounters with blossoms are already densely meaningful in seasonal and emotional terms. He learns about smells, and about the feel of flowers and nectar; he experiences his mother's desire and enthusiasm as she eats and, focused as he almost certainly is on her and what she is doing and experiencing, he knows that the two of them are surrounded by others who are doing much the same things. He is learning the twin joys of intimate experience and widely shared sociality.

Safely tucked beneath her wing as she forages, he is learning to recognise modes of sociality: nudging and pushing, becoming argumentative, sharing information, making decisions. The benefits for trees are a by-product of flying-fox etiquette and ethos (the embodied way of life); the more he learns about living in the flying-fox way, the more it is possible for symbiotic mutualisms to emerge from the encounters.

As night starts to fade it is time to go home. So, the little one already is learning about transitions between night and day, changes that will forever punctuate his experience of life. And he is learning about the return, gaining early certainty that there is a home and that it is not only possible, but indeed important, to return. He is not yet navigating, but he is tucked under the wing of an expert navigator, and so he is absorbing the fact that certain things, like finding home again, are knowable. As well, there is the early communication, the vocalisations that are unique to his mother and himself, and that distinguish the two of them from all others. I imagine that when they are back home they are replete, if it has been a good year for flowering trees. The mum is magnificently well fed, and the little one just keeps on suckling the milk. He knows her smell; he needs her gentle grooming, and he looks blissed out as she washes him with her tongue and hands. Their travel and foraging are finished for a while, and they gaze into each other's eyes, sharing their mutual experience of deep emotions. With every touch and every vocalisation, they are implicitly addressing those fundamental questions of who they are, what they do, what their world consists of and how they belong.

After a few weeks, my imaginary friend leaves her baby at home with other little ones. Rather than going out at night, he hangs out with other youngsters and starts to learn about life beyond his mum. He continues to absorb the wider ethos: bodies in close contact; poking and shoving while holding on to their own branch space; communicating with others; and gaining a growing repertoire of skills. Later he will learn more specifically to navigate and will discover his own expertise in the joys of exploration and return. Like all the other youngsters, he will develop his own mental map, and in due course he and his cohort will start up the noisy, bickering, wildly intense encounters that in a few years will lead to procreation.

The elements of this life narrative touch only a few points that are readily observable by outsiders. As humans, we are able to imagine glimpses of flying-fox ethos through direct observation, and through scientific research including new technologies. The experiences of carers who engage with this ethos in their intimate encounters offer us close empathetic understandings. It is true that we will never know just what it is like to be a flying-fox; our ways of knowing and imagining can carry us across some of the borders, and beyond that is the unknown.[1] And so we know that their lives are powerful and joyful, that their culture is transmitted from generation to generation, and that they experience the most wonderful exuberance, along with deep determination. From more violent encounters, we know that their commitments are so strong that only torture will sway individuals from their objectives.

My thought experiment takes a further imaginative leap, going way beyond anything knowable. What if my imaginary friend were a philosophically minded flying-fox? I wonder if she might be asking questions that go into the heart of her world of life. Why do trees blossom and why are the blossoms so nourishing? What holds us aloft? Where does all this power and energy come from?

I would love to respond: 40 million years of co-evolved mutualisms uphold these encounters and seductions. Mothers and youngsters, year after year, ride waves of ancestral power – flying-fox ancestors, tree ancestors, soil ancestors, ancestors of all the creatures whose lives contribute to healthy trees, such as bugs and mistletoe; and these ancestors, so many more than one could list, are connected through mutualisms that nurture and ripple in waves of ancestral power. They are greater than we can actually imagine, except to say that there is no singular power. Life emerges and is held aloft, or held in the world, because of waves of multispecies, mutualistic ancestors.

In praise of transformations

Back to Paris! The philosopher Lev Shestov was born in Russia in 1866; after the revolution he emigrated to Paris where he wrote and taught, becoming a key figure in the existential philosophy of his time. He died in 1938, leaving a legacy of impassioned essays in which he developed a critique of the West's commitment to universal progress, certainty and the idea that the mind of man could and should encompass everything. His analysis showed that mainstream western philosophy for more than 2,000 years had hyper-valued that which does not change, aiming to look beyond life so as to discover 'the eternal and immutable principles which govern the universe'.[2] A love for the immutable became, in west-ern philosophy, a lack of love for, and indeed a denigration of, the ephemeral world in which change is the condition of life. Thus, time and change were de-valued, and within the logic of denigration the world itself in all its changing unpredictabilities was de-valued.[3] He urged us (the West) to take birth more seriously, as part of the flow of life, and thus to value the fact that we are caught up in time and uncertainty. His existential stance led him to bear witness to life's unpredictability. Shestov urged people to embrace, rather than seek to evade, the challenges of living in a world of flux. Central to the question of what to value and what to dismiss was death. Shestov wrote: 'those who have meditated on this question [of the eternal] have established . . . so strict a bond between the idea of death and the idea of change that the two ideas at present are only one. That which changes now appears as insignificant, as miserable, as that which is condemned to die.'[4]

Shestov was prescient in his understanding that we of the West deeply need to understand that the brilliance and beauty of life lies precisely in its uncertainty and vitality. We are not promised immunity from vicissitudes, but, far more powerfully, we have the opportunity to see the world in its abundance. This proposition is challenging, perhaps especially when the issue is death; within mainstream western philosophy it invites a kind of glorious madness. And so, stepping away from the straitjacket of the unchanging eternal, we may be lucky enough to fall into love with all that is of this earth, all its transience, its fragile beauty, its unpredictabilities and uncertainties, and its countless joys and sorrows.[5]

We do live in a world of flux, and one form of evidence is found in the mainstream paradigm shifts now underway that challenge the West's attachment to the eternal and immutable. Uncertainty, probability, chaos and complexity are modes of understanding of how the biosphere really is, and are significant not only in science, as in quantum theory, for example, but in philosophy as well.[6] And yet, Val Plumwood reminds us, modernity has failed to give a life-affirming account of death.[7] A great many attempts to offer a loving account of death struggle under the weight of this long history during which death has been understood as the enemy. To step away from giving death more (or less) than it is due means embracing more fully the logic of both-and, thereby including death in the patterns and pulses of life on Earth. Death only exists because of life, and so to keep faith with life is to keep faith with the relationships between death and life.

Action

'Everything comes up out of the ground – language, people, emu, kangaroo, grass. That's Law.'[8] Hobbles Danaiyarri, an eloquent Yarralin Lawman, articulated his people's commitment to creation.[9]

His words imply much more than they explicitly state. There they were, the Dreamings, contained within the earth, and yet they experienced a desire that we understand retrospectively by considering their actions: they wanted others. It seems that Dreaming creators wanted life to come forth, to become active, and to flourish. They created diversity, complexity, connectivities and action. Creation sets up one of the foundational dynamics of life: the creators' mode of enduring potential (contained within the earth) works with the flowing of a multitude of ephemeral creatures who come forth and who carry life from generation to generation. This dynamic interface between the enduring and the ephemeral generates the life in which we are participants. Individuals are born, they die, and new generations take over. To be part of the world of ephemeral life is to face death as well as to celebrate the joys and beauties of becoming alive. Creation as process is twofold: the enduring potential and the ephemeral flow. Together they constitute a great participatory affirmation of ongoing creation.

* * *

Dreamings were not alone in their desire for others. Across a continuum of creation accounts in many times and places, numerous traditions tell or imply a similar story. G-d's breath upon the water, for example, was followed by action. Not satisfied simply to ripple the water, G-d started organising and making things. Here too we see the dynamic between enduring (eternal) and ephemeral, and here we see G-d's participatory affirmation, Yes! Both Dreamings and G-d, along with creators in other traditions, have within themselves the power of life; and from their action we see that their desire is for embodied, ephemeral beings. We see, too, that that desire is not to deplete the sources, but to include the coming forth as one of the great pulses of life. And so creators exist in at least two modes: in their enduring potential and in the ephemeral lives of transient beings who come forth but do not stay (forever).

The wonderfully poetic voice of creation/creators contrasts with the often dry voice of science, but they tell similar stories. The challenge of 'thinking with multiplicity' does not, indeed cannot, require us to take up either-or logic by choosing and rejecting, or to develop a hierarchy that grants greater legitimacy to one kind of voice over another. But it does ask us to think within the frisson of proximity and to take up the logic of both-and.[10] Contemporary paradigm-shifting science tells us that the biosphere is self-creating. 'Our planetary biota has had a regulatory effect on the chemical composition of the atmosphere, on Earth's global mean tempera-ture, on the salinity and pH of the oceans, and on other would-be solely physical factors of our global environment.'[11] The effect is that life itself has been working to create the biosphere so that it is favourable towards its own thriving. These circular, synergistic, life-giving relationships are thought by some scientists to call for a concept of 'active fitting', meaning that biosphere's parts are pat-terned and (in the long term) synergistic.[12] Biosphere science is secured in the West's primary domain of legitimacy, science, and one of the key points is that the biosphere has desires. Whether we think of origins as the work of creators or as the emergent work of life itself, the dynamic between an enduring potential and the ephemeral holds good. The 'flame of life' arises from these bio-sphere desires in modes of diversity, complexity and abundance, which, taken together, 'form the matrix of Earth's life-generating creativity and of the biosphere's robustness'.[13] In the long record of the biosphere, the proliferation of diversity brings novelty and adaptive fit to earth life.

At the heart of life is desire: the creators' desire for ephemeral life; living beings' desire for more life. This desire, this powerhouse of life's capacity to keep on coming forth, is fundamentally mutual-istic. The enduring and the ephemeral form a symbiotic agreement, to use Isabelle Stengers's words, and it is through the dynamics of this agreement that they produce 'new, immanent modes of exist-ence'.[14] The symbiotic agreement, let us recall, is complementary and dynamic, but not hierarchical. Each party in the dynamic has an interest, 'if it is to continue its existence, in seeing the other maintain its existence'.[15]

The complexities of life, while wondrously dense, are also immensely vulnerable because the ongoing-ness of life, its connectivities and co-becomings, happen through the actions of ephemeral living beings. This means that life doesn't just happen to happen, and it doesn't continue to come forth like an automaton. Ephemeral beings are crucial actors in the work of creation.

When death came to town

As with the enduring and the ephemeral, so with life and death. Death is not a stand-alone event; it only exists because of life. One of the crucial distinctions within the world of life is between those who die of necessity (eukaryotes, or mortals), and those who may live forever (prokaryotes) provided bad luck does not wipe them out. We mortals are those for whom programmed death is pervasive: cells die, individuals die, and species, too, are destined to die. The complementary dynamic is that sex is the partner to death. Mortals reproduce, thereby enabling the emergence of diversity and abundance. With sex, we mortals pass on some of our genetic make-up to our offspring; we do not live forever in the way of prokaryotes, but at the same time 'new non-copy organisms are brought into being'.[16] Diversity takes precedence over stasis.

Hans Jonas wrote an insightful essay on the burden and blessing of mortality. The burden, in his view, is like this: although we know we are going to die, we don't know when. Death could take us at any moment, and there is nothing we can do about that. We live always on the knife-edge. At the same time, the blessing of mortality is that we actually do die. His analysis works with the individual. If you were uniquely able to live forever, or for centuries, would you want to? Would you want to live, growing older and older, and finding all that you know and love dying around you? Or, if it were the case that all humans, indeed, all mortals, could live forever, or for centuries and centuries, where would we put them all? What would happen to earth life if mortals kept accumulating? The blessing of mortality, according to Jonas, is that it averts these ugly scenarios.

The story of mortality is equally the story of natality, as Shestov wanted us to remember and learn to love. The sometimes shocking fact of natality is that although each individual is in some sense 'new', the world into which one is born has long existed. In Jonas's words: 'natality ... means that each of us had a beginning when others already had long been here'.[17] The ephemeral world into which we are born is already a given fact of natality, and so we join the flows as beneficiaries of much that precedes us and will outlive us. Jonas tells us that the connection between natality and mortality is that the organically programmed death of parent generations makes room for their offspring. He is not talking about sacrifice, but about necessity. There is nothing in the great kingdom of mortal life that does not come into this world through waves of generations involving both death and birth.

Every day, and every moment, we draw life from our surrounds, and we avert death. Metabolism, Jonas wrote, 'is a continued reclaiming of life, ever reasserting the value of being against its lapse into nothingness'.[18] Life holds onto itself only through connectivities, that is through flows of energy and information that cross individual and species borders. As it is always coming, so it is always becoming, indeed, co-becoming. Life is always 'with others', and always in motion.

* * *

Jonas was working primarily with an either-or logic: *either* life in its vivacity *or* death as a kind of nothingness. When he and others spoke of precarity – mortal suffering and oblivion – they offer the kind of resistance to death that has been so prominent in the secular West. Jonas found beauty in death, as well as practicality. He wrote that through the work of staying alive, 'life says yes to itself'.[19] This affirmation gestures beyond dualisms, inviting us to think of death in both-and relationships: death as part of the dynamic between the enduring and the ephemeral. But still there is more: a participatory embrace of death would see natality and mortality as pulses that join individuals to ancestors, and that join ancestors to waves of power that carry life through the living and the dying. So this dynamic, too, depends on flux and metamorphosis.[20] The meaning of becoming-ancestral goes beyond what we can readily know, but that seems to include a particular ethical call: to die, to become ancestral, and thus to join the waves that bring forth life.

From my hospital window I can't help but notice the busyness of city life, but a horizon is drawing ever closer, and I am drawn to stories that take me to a threshold not yet fully imaginable. Old Tim Yilngayarri, one of the oldest Lawmen of Yarralin, had been a clever man when he was young, a keeper and practitioner of knowledge and action that went way beyond ordinary. He said that he had lost a lot of his power, but one of the gifts remaining to him was the ability to track the condition of a person who was near death. Sometimes he could follow the person into a zone of transition and bring them back. Other times he knew the person was going into a place where death was the only outcome. Old Tim's work with the person at the very edge tells us about that place where a living being is slipping: not yet dead but not able to come back. This threshold is a transition zone where the living and the dying have not quite lost touch with each other. Over there, some new pulse is soon to be released, but here, at this edge of unbecoming, living creatures are being pulled out of the life they were born to. A surge of life is returning to join that wave. In the transition zone pulses are being shaped. Death is imminent but has not yet arrived.

Natality

The dynamics of becoming and un-becoming form the context for the living experience of what the physicist and philosopher Karen Barad calls performative metaphysics. Briefly stated, her theory is that reality is ablaze with action and meaning. There is no such thing as passive matter or passive life. Every event is the product of that which preceded it and enabled it. Furthermore, every event engages in reality-making because each event is producing grounds for the future. The term performativity brings meaning-making action to the fore. We are all caught up in making reality, just as we are all made by it. Reality is not 'that which is', in some abstract or absolute sense. Rather, reality is that which is happening through dynamics of action and meaning.[21]

Because these dynamics depend on the past and are open towards the future, Barad calls them performative. Mutualisms such as those between flying-foxes and Eucalypts are nodes of 'intra-action' and thus are conjunctions of historicity and performativity.[22] The term intra-action is meant to remind us that action is transformative: it is not that two or more entities encounter each other and act to or with each other; rather, through action they shape and inspire each other, shaping the waves of power that carry them all. Thus, while historicity entails the fact that all life arises from that which preceded it, performativity ensures that historicity does not mean predetermination. The encounters of actors (also known as agents) are ripples of coming forth, and it is not known in advance just how they will materialise. Uncertainty and vitality are partners in life. We see this clearly in thinking about intra-action, or world-making.[23] The point is that worlds are both given and made. We always come after others, and thus come into worlds already given. At the same time, we are participants in worlding; we participate in shaping the worlds we inhabit and hand over to new generations.

The condition of mortality and natality is, moreover, a story of witness. This is our legacy and our responsibility: our very existence bears witness to our ancestors and to our surrounds. Our forebears were nourished, and they nourished others. And so we bear witness to all the others who made their lives possible. The metabolic imperative reveals our position as participants in entangled co-becomings: nothing stands alone, everything, at pretty well every scale, depends on others, all of us are borne on waves of ancestral power.[24]

It may be tempting to take for granted the work of life signalled by the unevocative term metabolism. Looked at from an even more detached perspective, this work commands our attention because it goes directly to the heart of what life does with Earth and biosphere. The research shift towards uncertainty and related modes of co-becoming shows that Earth life is a non-equilibrium system. Spiralling reciprocities in our biosphere channel energy into renewal, or, into order emerging against entropy. Entropy is the term that tells us that everything moves towards disorder. The partner word is negentropy, which is what life does: it resists disorder. Life – the 'biological order on earth'[25] – draws order out of disorder, organisation out of disorganisation. Connectivities are Earth's localised reversals of entropy. I find these matters quite exciting, but they may sound dry. Remember Eucalypts and flyingfoxes and how their mutual needs and desires draw them into cobecomings that resist entropy? It is exactly this: life resists entropy and in so doing attracts our fascination, inspires our awe, and wakes us up, yet again, to shared life.

Thom van Dooren, a key figure in the environmental humanities, writes about the incredible work that two albatross parents undertake each year to raise their one chick, flying thousands of kilometres across the ocean to feed on fish, and then to return home to their nests on a small island and regurgitate the fish to the waiting chick. Van Dooren reminds us that the parents are not committed to an abstraction such as a species; rather, parents work for their specific young. Furthermore, van Dooren, tells us, 'What is tied together is not "the past" and "the future" as abstract temporal horizons, but real embodied generations – ancestors and descendants – in rich but imperfect relationships of inheritance, nourishment, and care.'[26] Intergenerational flows are multispecies flows, and the surrounds that sustain these flows are both organic and inorganic. Albatrosses require fish; fish require oceans; oceans, like all water on Earth, move through a range of circulating loops at varying scales. At the largest of these scales: no water means no life.[27] So, to return to the immediacies of life, death and generations, we are multispecies becomings, the beneficiaries of all that nourishes us (organic and inorganic) and all that nourished our forebears. We are the beneficiaries of flows that so vastly precede us, and so widely connect us to other creatures and to the biosphere, that our indebtedness is impossible to imagine. Van Dooren expresses it beautifully. We are members of 'vast intergenerational lineages, interwoven in rich patterns of co-becoming-with others'.[28] We creatures who are alive today, human beings and psyllids, flying-foxes and Eucalypts, all of us, are pulses of life-becoming-with-others in webs of complexity, diversity and abundance.

* * *

James Hatley makes the point that there is no autonomous right to have been brought into existence.[29] We are not here because of entitlement; we are here thanks to the lives of others. Hatley refers to this condition as 'coming after' – we all inevitably come after our ancestors. Much of his analysis on witnessing arises from his study of human suffering caused by man-made mass death.[30] In carrying his analysis into multispecies waves of power my purpose is to broaden, but never to cheapen, the analysis; at the same time, one needs to be mindful of the fact that life for flying-foxes, and indeed for forests and many others, is situated at the edge of extinction. Ecocide is another form of man-made mass death. A focus on the goodness of life within this shadow of extreme death brings us into tough questions about the dynamics of the eternal and the ephemeral.

For Hatley, the relationships between generations must be understood in terms that include intergenerational narratives. Creatures come forth as part of the waves of ancestral power, and the generation that is born is not disconnected from the one that brought it forth. 'One is addressed by the lives one inherits. These lives inspire one, literally, breathe into one one's own possibility of existence.'[31] In the context of multispecies intergenerational waves, and from the perspective of birth, inspired narratives are part of the work that ensures that the young learn the parameters of their way of life, the ethos of their kind, so that they are culturally and socially equipped to go forth into their own participatory life which will be lived with and for others and will in its turn bring forth new generations. These flows are integral both to the lives of those who inherit them and to those who transmit them. Inspired narratives are central to the continuities of life, and they form bounded sequences. Any given group or population is shaped by its shared narrative, and thus 'can be seen as a wave of memory, insight, and expectation coursing through time, a wave that lifts up and sustains the individuals of each succeeding generation, even as those individuals make their own particular contributions to and modifications of that wave'.[32]

We can think of each wave as a pulsing transition from birth to life, from life to death. A pulse is the movement back and forth between the enduring and the ephemeral. On the face of it, a pulse is an absence followed by a presence followed by an absence, and so on, but in this ontological-ecological terrain of waves and intra-actions, pulses must be understood as dynamic transformations between the potential and its realisation. As sequence: quiescent followed by coming forth, followed by a return to quiescence. In the context of death: natality – life – death – ancestral waves. Each pulse brings forth more life.

An inspired pulse neatly references living bodies. There are pulses of heartbeat and of breath. Moreover, the term inspired relates to breathing. The body's inspired pulse breathes in, absorbs oxygen, and exhales. This is life. A pulsing vitality arrives with breath; it is our first cry, and perhaps our last cry as well. Breathe in, breathe out.

Desires

Something wants to move. Indeed, everything wants to move. Connectivity is essential to life, and motion is essential to connectivity. And thus there is desire: not only the will towards self-realisation but mutualistically towards the realisation of others as well.[33] As with the Sun and Rain and Flying-foxes, the call to get up and get moving awakens already existing desires. As we have seen in the discussion of how life arrives on waves of multispecies interactions, the desire for self also involves a need and a desire for others. Flying-foxes and their co-dependent trees are not the only ones to be captured in intra-active symbiotic agreements that glisten with desire. Freya Mathews writes that 'all beings desire what other beings need them to desire'. She offers some lovely examples in addition to pollinators: bettongs desire truffles, and as they dig they aerate the forest soils; emus desire zamia nuts, and their digestive tract prepares the seeds for germination. The list could go on and on. Mathews offers a larger analysis to show that creatures are 'exquisitely attuned' to the needs and desires of others, and she concludes that these mutualisms constitute an earthly 'proto-ethic'.[34] Not all creatures at all times are enmeshed in mutualisms marked by reciprocal benefits; this kind of co-evolved fit requires time and can be massively disrupted. I am not so concerned about the specificities of the desires of all life; when I hold the focus on flying-foxes and their mutualists, I find, as does Mathews, that there is a pervasive beckoning towards desire, and that the complexities of desire work with the biosphere's own desire for diversity, abundance and complexity.

Pollinators and angiosperms tell this story with enormous verve. The great angiosperm explosion in the Cretaceous period, perhaps 120 million years ago, enabled a huge outpouring of diversity. The key was the pollinator relationship. The explosion of flowering plants was matched by an exploding diversity of insects, and the relationship was mutualistic right from the start. For insects there was sweet nectar and nutritious pollen. For the plants there was radically enhanced pollination. Prior to multispecies pollination dynamics, plants depended primarily on the inconsistencies of wind to scatter pollen. Pollinators carry pollen more widely and more consistently. Better pollination allowed for greater diversity, and at the same time brought a greater degree of mobility to plants.[35] In western thought, plants traditionally have been deemed to be lesser creatures when compared with animals. Aristotle, for example, formed the zoo-centric view that plants were lower on the hierarchy of living beings in part because they are sessile, or immobile.[36] This characterisation focuses on individuals, and primarily on the life above ground. Looked at more generously, an angiosperm's inspired narrative would probably emphasise mobility as it is achieved across generations, for it seems that many plants, too, share in the widespread desire for mobility. Pollinators are the intra-active mutualists, and the more widely and consistently the pollen is spread, the greater the plant's opportunities for new generations to take up life at some distance from their parent and with greater possibilities for diversity and resilience.

With the explosion of animal-pollinators, desires and needs formed a circular and complex set of mutualisms.[37] The desire was *for* others who would do this work, and the opportunity they offered was to *enhance* the lives of pollinators, thereby offering compelling reasons for pollinators to visit. There is a desire for attraction: plants sought to capture the attention of pollinators, to invoke their interest and draw them into action. The result was multispecies seduction. Pollination had become a need: it is a way of life, a mode of achieving diversity, complexity, abundance and long-term resilience. Studies of plants' methods of attracting others show an astonishing range of colour, including colour in ranges humans can't see,[38] scent and other airborne chemical seductions, timing of flowering and nectar production, and more. Plants offer opportunities for others with their nectar and pollen, some of which is outstandingly nutritious. In this glorious co-becoming of mutual delight, plants and pollinators have produced a worldwide, multispecies festival of colour, scent and exuberance, and the vivacity of life has taken on whole new dimensions. Most of the foods that sustain animals are angiosperms, and humans are among the great beneficiaries in this metabolic story. Co-evolved loops of need, desire, opportunity and extravagance are the foundation of flying-foxes, plants and all the liveliness that ripples from their interactions.

I find it mysteriously compelling that so much of what plants put forth to seduce nonhuman pollinators is seductive to humans as well: the scents of flowers, their colours and shapes, their timing. As is the case with many other manifestations of life, ancestral power is beautiful. There is the beauty of form that comes from within living things: bilateral symmetry; proportionality; the fit between form and function.[39] Life in its pulsing waves produces these 'endless forms most beautiful', as Charles Darwin called them.[40]

And there is the beauty of intra-active patterns: the waves and ripples, circularities, repetitions and unexpected flourishes. Meta-beauty arises within all this *intersecting* beauty. Ancestral power is radiant.

Shimmer

'Shimmer' translates an Aboriginal term with the potential to awaken us. It refers to manifestations of Ancestral power. As an intra-active phenomenon, shimmer draws people to experience the dynamics of participation in the enduring and the ephemeral, birth, life and death, and mutualistic, multispecies waves of ancestral power. In a classic essay titled 'From Dull to Brilliant', anthropologist Howard Morphy discusses the art of the Yolngu people in the Arnhem Land region of North Australia.[41] His focus is on the Yolngu term *bir'yun* which translates as shimmering or brilliant. This is an aesthetic with corollaries in many parts of Australia; it arises in ritual, dance and song, and pervades many aspects of life more widely. I am using the term aesthetics in a non-technical way to discuss things that appeal to the senses; things that evoke or capture feelings and responses.[42] Aesthetics, in brief, includes the capacity to elicit affect. Morphy holds his analysis on the aesthetics and aims of Yolngu visual artists. He describes shimmer or brilliance (both terms are good) as the result of Yolngu painting which starts off with a rough blocking out of the shapes. In this first rough phase, the painting is described as 'dull'. We can understand this term as a description of the look of the painting just at the beginning. There is as yet no strong work; nevertheless, this phase is not happenstance. Transformations require this dull base so that change can be effected. The artist continues to paint, marking the shapes more clearly, and carefully placing the in-fill using clean, clear lines that constitute specific clan-based ancestral

Dreaming designs, and gradually a painting comes into vision. In its complexity it has become brilliant. A pulse has been initiated: something new has come forth; something has been transformed. Significantly, when paintings are made for ritual contexts, they are destroyed at the completion of the ritual. That which came forth has been transformed into brilliance and then is brought back into a state of quiescence. The making of a ritual therefore enacts pulses between dull and brilliant, offering participation, in the sense of world-making, in the flow of ancestral power. A pulse is performed first by bringing the painting from dull to brilliant. Then, at the end of the ritual, the brilliant paintings are rubbed out and made dull again, and another transformation is effected.

The making of a painting or a ritual enacts movement from quiescence to brilliance, and thus participates in the flow of ancestral power. It is performative in the sense of being world-making, and in the context of ancestral power the performativity enhances all sides of the intra-actions.

Every painting communicates beyond itself. Each figure and each specific style of in-fill references Dreaming creation. These are clan designs and clan Dreamings, and they connect the clan and the painting with the wider world. In the manner of connectivities, each Dreaming connects with other aspects of creation – mermaid with water, kangaroo with dry land, as examples – so the painting opens to country and creation, inviting the viewer also to open to life beyond the painting. With its referencing of Dreaming creation, shimmer always has multispecies connectivities. In communicating beyond itself, the shimmer of careful and complex work captures the eyes of those who look at it. *Bir'yun* is a kind of motion, artists say. Shimmer grabs you. It allows you, or brings you, into the experience of being part of the vibrant and vibrating world. When a painting reaches brilliance, for example, people say that it captures the eye much in the way that the eye is captured by sun glinting on water. There is power here; Yolngu people compare the glint of brilliance with 'the flash of anger in a shark's eye'. There are similar captures all over the place: water capturing and reflecting the sun, the sun glinting on wet leaves, the eyes of the beholders captured and enraptured, the shark flashing its look, the rippling intra-activity of it all.

This glittery shine, this shimmer, has a twofold effect for Yolngu people. It testifies to the presence of ancestral power, and it may create in the person who is captured by it the actual experience of that power. When one is captured by shimmer one experiences not only the joy of the experiential capture, but also, and more elegantly as one becomes more knowledgeable, one participates actively in the flow of ancestral power.

Morphy was fascinated by the fact that while the full meaning of bir'yun is culture-specific, the experience of shimmer is cross-cultural: we are all capable of being captured by shimmer, and we are all capable of knowing that something significant is happening. To this we must add that the term does not distinguish between domains of nature and culture, a distinction that is largely irrelevant in Aboriginal culture. Clan designs and clan Dreamings call the world into a specific style of in-fill, and each figure communicates itself and its nonhuman relatives. The making of a painting or a ritual that enacts movement from quiescent to brilliant thus participates in the flow of ancestral power. It is performative in the sense of being world-making, and in the context of ancestral power performativity enhances all contributors to the intra-action.[43]

The dynamics of quiescent-to-brilliant pulses replicate across contexts and scales, recursively and iridescently. But then we leave the world of art and enter into a more widespread consideration of pulses of shimmer. Artists want the painting to be dull in order to emphasise the transformation, but in the world of living things, pulses are not dull. Rather, the transformation from quiescent to brilliant is a process that effects transformations in both directions: the coming forth, and the return to a certain quiescence that signals the potential to come forth. Consider the seasons: from the desiccated, browned-off country at the end of the dry season, to the brilliance of lightning and the glint of sun on raindrops, to the green springing up of new plants, to the birth of new animals – all this abundant coming forth happens within that rhythmic matrix of aesthetics and power – from quiescent to brilliant. Think for a moment of flowering trees. They have their time of shimmer, but it does not follow that they are otherwise dull. Rather, pulses rely on two moments: the coming forth and the withdrawal. Power manifests in the transformations from one moment to another. The pulse of shimmer captivates by transformations of actual coming forth. And so, vibrancy is always present – sometimes contained, sometimes bursting forth. Shimmer testifies to the pulses of earth-life: seasons, new births, new growth, sap moving, lightning strik-ing, the winds' action rippling the water and the sun glinting. The flow of life is arriving on waves of ancestral power, carried on the generations. The process of 'making manifest' is a transformation achieved performatively. There is nothing that does not participate in the flow of power;[44] this is the flow of life, all life, arriving on waves of ancestral power and singing it up into more coming forth.

Each flash of ancestral power is a site of witness, offering testi-mony to the real and calling for participation. In these dramas of withdrawal, transformation and emergence, shimmer spreads its wings and calls out: Yes!

Excursion into song

The power of ritual cannot be forced, it cannot be demanded, and whether it communicates or not depends in large part on who is paying attention. Aboriginal Australians are masters of the arts of ritual and transformation. I have had many opportunities to wit-ness this fact, and land claims were one such complicated context in which the power of ritual, understood differently by different parties, and perhaps not understood or respected by a few of the participants, nevertheless affected those who took notice.

Jasper Gorge, near Yarralin, was created by the Black-headed Python Dreaming. In the Jasper Gorge-Kidman Springs land claim, we made a bush court in an open area at the edge of the gorge. The flying-fox track was just above us in the high country, and a little mob of permanent residents was close to us in their camp in the pandanus trees. Bush hearings were complex events because two laws were coming together, trying to make sense of each other, and for many of the participants, trying to find a fair and empowering outcome. From the perspective of Anglo-Australian law, the participants were all human, but the Aboriginal participants were aware of a wider set of attentive and authoritative participants, including ancestors and Dreamings. Deep knowledge of country and creation was being shown and discussed; it was intensely serious stuff. The flying-foxes were significant participants throughout, but as the dreaming aspects of their law were held as men's sacred business, details were not part of the open evidence.

In this claim the women were disadvantaged: their most powerful evidence was secret-sacred for women only. And yet, the judge, barristers and anthropologists other than myself were all men. The women thought about allowing these outsider men to witness their ritual. On the morning of the day of their secret evidence the women went to the area where they had prepared a ceremony ground. They had brought the sacred objects from the keeping place, and their plan was to spend the day preparing for ceremony which would be witnessed in the late afternoon. Not long after we had assembled, the senior women told me there was a problem. They wanted me to go back to camp and convey the message that they had decided not to show the ceremony to men after all. They explained: 'From Dreaming right up till now no man has looked at this. We can't lose that Law.' They spent the rest of the day doing ceremony without men. The country and the Dreamings were witnesses, and so too were the non-Aboriginal women who had been invited. With all the men excluded, the women's ceremony had no value as legal evidence, except that I was able to note for the record that it had occurred.

Then, on the last day of the hearing, the traditional owners announced that they intended to put on a 'show' for everyone. There would be song, dance and body art; the purpose was to show the judge that women and men work together in the management of country and knowledge. This final 'show' was a ritual inversion of the hearing, and the reversals were so marked and humorous that the intent could not be missed. It was organised for night-time, it was held outside of the bough shade where the hearing had been held, and the spatial organisation was determined by the traditional owners. The judge and all other non-Aboriginal participants (lawyers, anthropologists, field assistants) were told that we would be summoned when the traditional owners were ready for us, and that we could just sit and wait in the meantime. When called, we were told to pick up our chairs and come forward to arrange ourselves in a row facing the area that had been designated as the dancing ground. The traditional owners used vehicle headlights as spotlights, and with imaginative use of space, lighting and personnel they created an event that included women, men and children. The children provided humour inadvertently (in part), and some of the men took up the clowning role, thereby indicating that the most serious work that evening was being done by women.

The women danced and sang. It was their first opportunity to show the judge that they had Law to back them up. This dance was not secret, and they were happy to have it witnessed by everyone. At the close of the event a senior regional Lawwoman grouped the women together to voice the unique call that speaks to country. With her hands and arms she gestured in downward motions, and then she spoke words that encompassed the whole of the claim – it was all there in her voice and her actions – the site visits that the judge and others had witnessed, the women's business that the judge and others had been excluded from, and the everyday evidence to which everyone had contributed. Turning to the audience she explained: 'We opened the country up for you, and now we are closing it down again.'[45]

What does open country look like, and what happens when it closes down? I am not an expert, but I can say this: when country has opened up, its ongoing waves of life become manifest. We had been given the opportunity to experience the shimmer of it, and when the Lawwoman closed it down again I think many of us experienced a sense of completion. There had been a great pulse of power actualised into this intra-active event with its multispecies participants and its abundance of calls and responses between the enduring and the ephemeral. When the Lawwoman closed down the country, she called it back into potential. The country was by no means dull, and for those who understood it and knew how to experience it in its fullness, it was rich in potential. No longer, though, did it glow in its extra-ordinary state of shimmer.

7   The Vortex

The biosphere has its own 'normal' states, punctuated by waves of death and bursts of life. Natality and mortality form a dynamic that is tilted towards natality and, hence, towards the proliferation of life in its abundance and diversity. Unforeseen and terrible events such as epidemics, volcanoes and many others drive periodic onslaughts of death, and as we learn more about 'natural disasters' we also learn about resilience and the emergence of new life.

Extinctions

As death is part of life, so, too are extinctions. Mass extinctions, however, are rare. In these events, huge numbers of diverse plant and animal species are wiped out in a relatively short time frame. The most recent such event was the mass extinction that caused the end of dinosaurs along with many other species. It is true that birds have continued some dinosaur lineages, in radically altered form, but there is no doubt that about 66 million years ago there was an enormous, catastrophic loss of life and diversity. The event was probably triggered by a meteor crashing into Earth, but that explanation is not conclusive.[1] The extinctions were followed by a massive flourishing of novelty and diversity; mammals and flowering plants are part of this diversity, as is much of life on Earth as we know it. The unexpected loss of existing species is counterbalanced by the evolutionary emergence of new species, and new species outnumber those that have disappeared.[2] In this way, life becomes more diverse through time. 'Diversity promotes diversity.' New species arise, and 'the more complex and interesting biological diversity seems to become'.[3] The flame of life burns brightly, one might say.

As is surely well known, today it is human beings who are the driving force that, directly and indirectly, is pushing the biosphere towards the dominance of death. Not all humans are equally implicated in this extinction event. The deathwork is being driven directly and indirectly by the actions of modern industrial societies and its effects are global. Because humans are so deeply involved in this current extinction event, it is best referred to as a man-made mass extinction. The rate of extinction now is at perhaps ten thousand times the background (or 'normal') rate.[4] As ecologist Stephan Harding says, 'we are hemorrhaging species'.[5] And yet, what is occurring is more dire than numbers indicate. In expanding domains of loss, death takes on new, cascading dynamics. Connectivities come undone, and feedback loops of destruction overtake the symbiotic agreements that promote life.

A botanist specialising in Central America wrote many years ago, 'what escapes the eye ... is a much more insidious kind of extinction: the extinction of ecological interactions'.[6] A good example of this functional extinction concerns a species or population that is no longer able to play a significant ecological role. When pollination stops, plants are unable to reproduce. Even now, whole forests may be places where no future generation will arise. From a functional point of view, many species are 'extinct' long before the population crashes, leading some botanists to describe them as the 'living dead'. Another expert on plants and pollinators writes: 'It will be tragic if the remaining natural areas of the world are filled with ageing plants silent as graveyards. . . .'[7] Similarly, a poster promoting understandings of flying-foxes as pollinators shows a flying-fox in a tree, and bears the slogan 'No me, No tree'.[8]

In addition to such direct loss, there are extinction cascades or vortexes in which connectivities unravel and mutualities falter. Some tree species are co-evolved with flying-foxes, as discussed, and the out-crossing pollination over distances provided by flying-foxes ensures variety and resilience within species. In Carol Booth's succinct words, fragmentation in human-dominated landscapes results in loss of genetic diversity. Over time, there will be a loss of trees 'capable of rapid response to climate change' or, even more drastically, such loss may 'eliminate the genetic variants for rapid response altogether'.[9] The dynamics of flying-fox deaths tell multispecies stories.[10] They are 'biocultural' stories, a shorthand term indicating that humans and nonhumans, engaged in relationships of simultaneously biological and cultural significance, are involved. Direct and indirect human factors intersect with the lives, needs and connectivities of animals and plants, and whole regions of multispecies worlding are poised to disappear.

* * *

In our current context, I call this era of man-made mass death a time of 'double death'.[11] It is made up of many small increments as well as great waves of devastation, and it impacts both on the balance between natality and mortality, and on the capacity of ecological systems to recover from devastating impacts. It is not that the whole biosphere is overtaken with death, but rather that death continues to pile up; renewal and resilience cannot keep pace.

Double death is a despoiler. It is an affront to the biosphere's flame of life and to our very being as creatures of Earth:

- so many losses occur that damaged ecosystems are unable to recuperate their diversity: the death of resilience and renewal, at least for a while;
- so many extinctions that the process of evolution is unable to keep up, and more species die than are coming into being: the death of evolution itself, at least for a while;
- the unmaking of mutualisms, the loss of symbiotic agreements;

- the unmaking of lineages, unravelling the work of generation upon generation of living beings; the death of temporal, fleshy, metabolic, narrated, inspired relationships across generations and species;
- the assault on Earth's shimmering waves of ancestral power.

Staying with trouble

So, we are living in a deeply troublesome time. Action against flying-foxes is situated within the current violence against nature. Two species of flying-foxes are listed as vulnerable to extinction under the Australian federal government's Environment Protection and Biodiversity Conservation Act 1999: (*P. poliocephalus*; *P. conspicillatus*) – Greys and Spekkies. The delightfully named Spekkies have a range that takes them beyond the borders of Australia, and on into Indonesia and Papua New Guinea. On the basis of this widespread population, the international organisation that monitors species' health (IUCN) does not list Spekkies as threatened. Only the Australian contingent is listed. Greys are listed both federally as well as with the IUCN; their range is limited and their demographic profile is in decline. The story of how two species of flying-foxes were tipped to the point of being declared threatened species is a story of the Anthropocene. All are biocultural, but not all human factors are directly intended to eliminate flying-foxes. Rather, the entangled consequences of multiple events, many of which may be accidental, form a dynamic whose outcomes include waves of death.

Donna Haraway advises that the intra-active commitments we share with others require us to 'stay with the trouble'. There is no guarantee of better outcomes; rather, commitments ensure that double death does not go unchallenged and that living beings are not abandoned. Equally it is a voice of witness, but not necessarily (indeed often not) a voice of triumph. One of the outstanding figures in this regard is Jenny Mclean, whose work with Spectacled flying-foxes in North Queensland demonstrates many of the multispecies dimensions of the biocultural dynamics of disaster,

and the enormity of a commitment to work on the side of those who
are endangered.

* * *

If you fly to the tropical resort town of Cairns, in the far north of
Queensland, you can rent a car and drive up the escarpment on
a winding road that takes you through dense, beautiful, tropical
rainforest. This is the place of some of the oldest rainforests in the
world, rich in diversity, and home to ancient Gondwanan species
that now live nowhere else. Spekkies play a critical role in rainforests, as they are unique in providing seed dispersal to a variety of
plants. Risk to Spekkies is bad news for places such as the World
Heritage Listed Daintree Rainforest.[12] When you reach the top of
the escarpment, you come out onto a region known as the Atherton
Tablelands. Suddenly you are in savannah country. There's no more
dense rainforest, it is all open woodlands with golden grass and
tall Eucalypts and Corymbias; or, it would be like that if there were
not so much land clearing. If you were to continue to travel west
you would follow the Savannah Way across the whole of North
Australia, remaining within that wide ecosystem all the way to the
Indian Ocean. Jenny Mclean's Tolga Bat Hospital is just at the point
where rainforest and savannah meet each other. The hospital is a
large, multipurpose, NGO complex situated on the five-acre block
of land that is Jenny's home. It is located near the outskirts of the
town of Atherton and thus occupies a site in the midst of rural
industries including agriculture and cattle.

Before leaping into stories of Jenny and the Bat Hospital, I need
to say that this region is a hot spot for human conflict about how to
live, or not live, with flying-foxes. Some of the most vicious public
animosity towards flying-foxes comes from this region. One of the
state MPs (Member of Parliament) 'has long held a pro-culling
stance on the animals'. He takes the view: 'We are all sick of the
stench and filth, and the devaluing of local properties near the flying
fox populations; people need to be prioritised over bats.'[13] In 1980,
well before the Bat Hospital was founded, people in this area were

responsible for a major attack on local flying-foxes: the Tolga Scrub Massacre.

> Early in 1980, Atherton Shire Council declared war on an esti-
> mated 40,000 flying-foxes roosting in the Tolga Scrub, blaming
> them for tree damage in this precious rainforest remnant. The
> environmental managers issued a 'permit to destroy' and council-
> lors took up their guns in what the media described as 'an act of
> appalling cruelty'. Reports of dying bats hanging from trees and
> scattered over the adjacent road caused a public outcry, resulting
> in a 7-day ceasefire and use of more humane scare tactics to
> drive the bats on to less sensitive roosting sites. It was over 6
> years before the bats returned . . . the real culprits [inflicting tree
> damage] are huge flocks of white cockatoos and the occasional
> large camp of Little Red Flying-foxes.[14]

One irony of this conflict is that the Tolga Scrub itself is endangered, being one of the last remnants of a critically endangered type
of rainforest. It is already a survivor of a century or more of land
clearing. Clearly it should be protected no matter who is damaging
the tiny remnant. The Bat Hospital is one of its strong advocates
and has been awarded grants to help conserve the area.[15]

In contrast to 'massacres' and calls for 'culling', the Cairns
Regional Council, 100 kms away by road, has an admirable history
of negotiating tolerance. The Regional Council relies on the federal listing of Spekkies to support its nuanced and evidence-based
approach to finding ways into sustainable multispecies communities. 'Cairns Regional Council is working closely with local and
national experts to determine a flying-fox management approach
that balances conservation of these important native animals, and
the amenity of residents, businesses and visitors.'[16] In 2018 the
Council revived the annual 'Cairns Bat Festival'.

My experience of flying-foxes in Cairns took a twist when I
became too unwell to carry out research; I went as a holiday visitor
and stayed at a resort hotel on a nearby beach. The outdoor restau-

rant area was set amongst massive paperbark trees that had been incorporated into the design of the place rather than chainsawed out of existence. Food was not served there in the evening out of concern for contamination by flying-foxes, and rightly so. Flying-foxes don't care where their poo goes, as long as it goes down. But we were perfectly willing to risk losing a cocktail, and we took our drinks outside to enjoy the sunset. At dusk the flying-foxes came too, happily foraging high above us in the flowering trees. This area had been cleared and developed decades ago, and I was impressed that any large trees and free-living animals remained. And yet, flying-foxes were there, and it was a rare delight to be part of a multispecies happy hour.

* * *

Jenny Mclean is immensely capable with flying-foxes, people, planning, building and education. She is an outstanding advocate and carer with a remarkable commitment to every project she takes on. Tolga Bat Hospital is the home of a not-for-profit organisation dedicated to both care and change: 'We are a community group that works for the conservation of bats and their habitat through rescue and landcare work, advocacy, education and research.'[17] In 1990 Jenny started a care programme operating out of her home. She has enlarged the place enormously, and the hospital is a care facility, an educational centre and a place where scientific research can be carried out. The home/hospital is set on a large block of land that slopes gradually down to a creek.

Adjacent to her home, Jenny has built the tallest flying-fox enclosure in Australia. It provides a generous flying area and allows for the creatures' desire to be high above the ground. There are a lot of resident flying-foxes; the enclosure enables them to enjoy every one of the factors Jenny identifies as essential to good quality of life for flying-foxes in long-term care:

- adequate housing – at least six meters high so that they can fly
- suitable social life – appropriate mix of age, gender and species

- adequate food
- adequate sunshine
- adequate mental stimulation – this will be provided by other flying-foxes, as long as the social composition of the group is right
- the ability to get away from humans – the high cage allows this; draped cloths can also provide privacy
- no stresses, such as no kids running around.[18]

Close to the large enclosure is an orphan cage, and separate from that is another enclosure for microbats. Jenny has developed a visitor centre with advanced ecotourism accreditation that also provides opportunity for people to 'adopt' a bat and support it through regular donations. More recently, Jenny added a large new building called the 'nursery'. It is a multipurpose space for caring for injured creatures, and for carrying out scientific research. Adjacent to Jenny's home are quarters for volunteers, and out the back there is a cold house for storing the huge amounts of food that are required.

Individuals in long-term care are clearly unable to live freely in the bush, and yet they are enjoying life. The burden of care is huge, but as Jenny says, 'this is the big dilemma. It's a lot of work and money, but otherwise all you can do is kill them.'[19] The life histories vary; each of these creatures has been rescued from some terrible fate – death by untreated electrocution, agonising suffering stretched out on barbed wires, for example. Two flying-foxes had lived appalling lives as a roadside attraction, held captive in a tiny cage and fed inappropriate food. Jenny said that carers had worked for a long time to secure the release of these two. Rusty and Chopper had lived so long in a one-metre cage that for a long time they simply didn't move more than a metre in any direction. They had lost their understanding of themselves as mobile creatures, and it took time for them to realise they could join the others, scrambling to the high reaches of the cage, flying, and coming back down for fresh fruit treats. Rusty was willing to allow himself to be handled, and became a great educator as well as a participant in scientific research:

Rusty is a very large male who came to us from a private wildlife park. Bat carers had been trying to 'rescue' him for a number of years, as his living conditions at the park were not very good. Once rescued, we were asked to take him and his mate Chopper. We tried to let them both live outside the cage but they were too institutionalised and kept wanting to come back in. Rusty's fur became dark and shiny as his diet and living conditions improved.

Rusty is very food-obsessed, and has become quite fat. He steals milk from the babies and special foods from the Little Reds. He tries to get food from the volunteers who clean the cages in the mornings. We have to watch him!

Like many Black Flying-foxes he has dark red or rusty coloured fur at the back of his neck, and hence his name.[20]

I became partial to Rusty because his burly black body and ochre ruff gave him the classic *P. alecto* look, and also because he had an unusual willingness to interact with humans. On a visit I made after the new nursery wing had been built, I observed Rusty's semi-willing participation in a heat stress research project that measured stress responses to different combinations of temperature and humidity. The research was undertaken with animal ethics approval, and Rusty was not subjected to anything more than mild discomfort. He seemed relatively relaxed in the heat box; his heart rate and temperature were being monitored, and most of the time he looked as comfortable as a flying-fox could look being confined in a tiny box with a thermometer in his bum. In spite of dislodging the thermometer a few times, Rusty was definitely making his contribution to science! Thinking about the lives of these two individuals brings Jenny's dilemma into sharp focus. Why on earth would anyone want to put them down, and yet how on earth could she cope if every flying-fox were given permanent refuge? What can be done with institutionalised individuals who simply can't cope with free living, and why should they, or anyone, die just when they've been saved and given an opportunity to enjoy life?

The work of care goes on every day, taking up most of the day, throughout the year. The floors of the enclosures must be cleaned. The detritus is shovelled into enormous worm farms that produce marvellous fertiliser for the gardens that are also a feature of the place. Fruit, kilos and kilos of it, must be cut and strung on wires that are hung from the top of the lower cage so that flying-foxes can come and eat whenever they choose. And the food must be sourced and fetched. If there are babies, they need to be fed and cuddled, weighed and measured. In the afternoons the hospital is open to visitors who, for a small fee, have access to certain areas. The information area is fascinating, but the outstanding part of the visit, for most visitors, is the rare opportunity to come face to face with flying-foxes under safe and controlled conditions. Visitors do not touch the creatures, but they can come close. It is as if a long-suppressed curiosity is unleashed, and people ask question after question, filling their minds with facts and offering the experience of wonder and appreciation.

Comments are mainly enthusiastic. For example:

School Group – Great Information, amazing, Fantastic! Very
    Impressive! Filed my curious mind with interest.
Eva – the bats bestilled my heart. ♡
Lea From France near Paris: the bats are so cute!
Les chauve souris sont trop mignons!
From Florida, USA: You all are an inspiration! Keep up the great
    work.
Brilliant! Call me if you need carers in Cairns.
From Holland: This visit, seeing the cute eyes has caused us to
    lose our fear and coming to like bats![21]

The hospital is also open to qualified research scientists and scholars such as myself:

We're getting more and more vet students. They're interested in
practical hands-on experience. We get involved with researchers

as much as possible, particularly in tick season, we try to find homes for all the dead bats. If they can be used in research for some purpose, at least they won't have died in vain. [And] We've trialled radio collars, satellite collars. We're lucky to have CSIRO here in the area. We get on very well.

I visited the hospital several times, staying for a few days each time in the volunteer accommodation and revelling in the opportunity to listen to the flying-foxes as I fell asleep and whenever I woke up. At night curlews sang their eerie, whistling calls, and by day the parrots added their voices to the full-time cadences of flying-fox sociality. I made sure I came at a quiet time so as not to interfere with the focused work in the time of intensive care.

The reason for Jenny's extensive volunteer space is that at a certain time of year there is a massive surge of flying-foxes in need of care. People come from all over the world for the privilege of volunteering with Jenny and the flying-foxes. The story of why such intensive care is needed shows a tangle of biocultural factors that contribute to pressing Spekkies towards the edge. The crucial period is October, November and December of each year: birthing time, and the time when babies are totally dependent on their mothers. Around 1990, carers in the region realised that an unusual number of unexplained deaths had a pattern, and in that year the probable cause was identified. Jenny came into the story in 1997 when the maternity camp in the Tolga Scrub near the hospital was affected by sickness and death. Mothers were dying, and babies were being orphaned at an alarming rate. The evidence is not fully conclusive, but it seems that the problem involves a combination of factors. In this part of the Atherton Tablelands there is an invasive weed, a wild tobacco from South America: *Solanum mauritianum*. This tobacco favours forest margin, scrub and cleared land; it is aggressive and fast-growing, and it is toxic to humans. The plants have berries and flying-foxes forage on these berries during their crucial maternity season. It may be that they have turned to this low-growing food because native foods are lacking. The berries themselves, though, are not the problem. Rather, in this multi-species knot of trauma, ticks infest the tobacco plants; evidence shows they may be more prevalent on this plant than on many native plants. In any case, the tick (*Ixodes holoyclus*) paralyses its host. It takes a few days for the paralysis to set in, but when it does it is usually fatal to flying-foxes. Although the tobacco is new, the ticks are not. Ground-dwelling native animals have lived with ticks for a long time and have become immune to the toxin. Flying-foxes, being arboreal, never developed immunity.[22] Only recently have they started foraging close to the ground where they pick up ticks and fly back to camp; females are able to continue to nurse their young for a short while, but then they become weak and die. As they weaken, they fall from the tree. If the baby is still attached to the mum, both of them fall. The mother dies, but the baby is still alive, calling out for its mum to respond. Other babies, whose mums go out foraging and weaken or die before they are able to return, cry in the trees, hoping they will come back.

In the midst of this suffering, Jenny and her team of dedicated volunteers walk the forest looking and listening for flying-foxes. Jenny took me for a walk in the Tolga Scrub to show me the terrain she and the volunteers traversed in their rescue work. I was mindful of the warnings about this and other tropical rainforests: 'The volunteer needs to be tolerant of large spiders and other wildlife as well as recognise dangerous plants.'[23] It is not work for the faint-hearted at any level. Some days the team picks up as many as fifty individuals, living and dead. Some adults can be saved but many cannot; many babies are orphaned either because the mother is already dead or is too far gone to be saved. In the most stressful tick seasons Jenny has had 500 orphans, although 200 is more usual, and there could be another 100 or more adults affected by the toxin. The babies are at different stages of development; some can graduate to solid foods quickly, others need to be bottle-fed for longer periods. Jenny described a 'conveyor belt' style of organisation, as a volunteer who was in the orphan room cleaned babies, wrapped them in a blanket and placed them in a box. When the box was full, the carer brought it outside to another group seated on the veranda. Each person took a baby and fed it from its little bottle. The fed babies went back in the box to be cleaned and wrapped again and then left to sleep, while another box was brought out and the carers each took a baby and a bottle, and so it went on hour after hour, day after day. The aim was to get the babies onto solid food as quickly as possible. 'It's pretty full-on. It's great if you've got a nice number of people, it's very sociable and it happens relatively quickly. If you've got too few people, it just goes on and on, and by the time you've finished, it's time to start all over again.'[24] In addition to caring for all these orphans and sick adults, Jenny and her team keep notes. The quality is excellent, like everything Jenny does, and they form the basis of the first detailed scientific study of the tick-tobacco-flying-fox nexus.

Jenny's preparations for the annual influx include getting in a bobcat with a big auger to dig a number of deep pits where the dead can be buried. Flying-foxes who cannot be saved, and there are many of them, are euthanased and buried along with the dead bodies that are brought back. It is important not to leave dead bodies in the bush because the flies build up. 'The bats get attacked by flies as soon as they start to show signs of not being well. We got lots of babies coming in all covered with maggots. The more we do to minimise the number of flies, the better. It's easy if you've got these big holes.' The holes must be carefully placed to be away from the water, and they must be deep. If an animal were to dig into a pit and eat a creature that had been euthanased, it could die too.[25]

The aim of saving lives is to get the individuals back to the bush. Jenny has a release site nearby where individuals can make the transition back into the way of life they lost when they were struck down. Every release individual is microchipped, as are most of the individuals in care.[26] Research projects gain data from micro-chipped flying-foxes, but knowing who's who in the region rarely provides good news since the information is most likely to be noted if the individual dies or comes to harm. Life in the bush is full of hazards. Some of the rescued individuals, newly microchipped and released, go out to forage, return to the tobacco plants, eat berries, get a tick, slide towards paralysis, and (if they are lucky) end up back in care. Life in the bush also exposes flying-foxes to the risks of human hatred, vilification, torment and killing. As new legislation enables greater numbers of direct and indirect deaths of flying-foxes, Jenny's work necessarily engages with the grim dynamic of love versus hate played out in the lives and deaths of flying-foxes.

Jenny is a fifth generation North Queenslander. Her ancestors settled this area, chopping down forests for timber and the planting of sugar cane, displacing and killing native animals and dispossessing Indigenous people. In short, they were active participants in the war against natives and nature that characterises so much of the formation of settler societies such as Canada, Australia and the USA. Hesitantly, and with deep humility, she spoke of how her efforts may work against that history. In this sense, she is engaged in a strong act of decolonisation. Reparative work in dialogue with the past is deeply significant and consistent with her immediate purpose of sustaining all those that have a chance of survival, and of returning them to their forest homes as soon as they are ready for release. Back in the bush they are again subject to the hatred, vilification and killing.

When I asked Jenny why she does this work, she spoke of simply being there: this suffering was going on around her, she was there, and so she responded. Being an extraordinarily practical person, she responded by building a hospital, setting up a non-profit organisation, organising volunteers, offering education programmes, and finding ways to ensure that it will continue when she is gone.

The Tolga Bat Hospital works with all these tasks, relying on the support of other carers, and large commitments of time and energy on the part of volunteers. It is a centre of multispecies alliances, some of which are enduring and others episodic; it truly embodies and enacts the challenges, successes and precarities of multispecies cooperation in this time of peril.

The pteropucidal black hole

Greys inhabit the most populous (and therefore most perilous) part of Australia, but at the same time they are better studied than the others. Two extremely interesting scientists, Leonard Martin and Allen McIlwee, undertook an analysis of the population dynamics of Greys (in 2002) when the profile towards extinction was already discernible. A more recent study of reproduction and longevity shows that the situation for Greys is worse than Martin and McIlwee realised when they did their analysis. The more recent study showed that the numbers of Greys was rapidly declining and would lead to extinction in about eighty years.[27]

Martin and McIlwee were concerned about the impacts of contemporary 'modern mortality' on creatures whose ancient evolutionary history had taken place in a more benign environment. They examined the classic population factors, including number of offspring (low) and longevity (relatively long). In addition, they factored in the traditional uncertainties of life in Australia, notably heat stress and starvation. They noted, as do others, that predators were not a major factor. Their conclusion was that the low reproductive rate was well adapted to the environment in which flying-foxes had lived for so long.[28] What changed was the arrival of Europeans with their 'habitat destruction, persecution and culling'.[29] I will return to direct human actions, but first let us consider how ancient factors such as heat stress and starvation are now also entangled in biocultural dynamics that also involve humans.

We saw with heat stress that habitat clearance was an issue. On the one hand humans were building their homes in areas that attracted cooling breezes. On the other hand, flying-foxes were being driven out of areas where they had once been safe. The Melbourne Botanic Garden expulsion forced flying-foxes to leave a place where they had not experienced heat death; in one of their major new sites, they experienced massive death. In the year of the expulsion there were no juveniles, and heat stress took the lives of many more. Similarly, in the 2013 Sydney heatwaves, 15,000 flying-foxes died over the span of a few days. The passionate carer Storm Stanford pointed towards the growing scale of the disaster: 'Ninety-nine per cent of the animals that are dying are juveniles. We are losing a significant proportion of the next breeding generation.'[30]

Clearly, to lose one year's reproductive capacity is to cut into the dynamics of survival. The larger problem is that it happens again and again. Spontaneous abortion is another factor. There are reports of pregnant mothers suffering abortions en masse. An incident in 1978 was described this way: 'there was a mass abortion of close to full-term foetuses of the grey-headed flying fox. Several thousand females (out of approximately 12 000) lost their young which were expelled complete with the placenta and festooned the branches of the understory of the camp. Crows and foxes were feeding on the carcasses.'[31] Possible causes include bacteria, parasites or disease. Similar events are documented from other places. Stress is known to be detrimental to pregnant and lactating females, and the young are vulnerable to human violence and to heat stress, as discussed.

Starvation involves a similar story of habitat loss and climate change in relation to food rather than heat. Whether across small or large regions, loss of food is most lethal for juveniles. They may die outright, or mothers may abort spontaneously in the face of starvation. Louise Saunders described one such event caused by humans who were trying to drive away flying-foxes by cutting down trees, including trees that bore food.

The mothers ran out of milk and then the babies died of starvation. And . . . the mothers hung onto them . . . until they just couldn't hang onto them any longer. The poor little things were just bags of bones. But it's that suffering . . . Who'd wish that on an animal? And yet, they [the humans] were celebrating and condoning it.[32]

In 2006 the Bat Rescue online newsletter carried an article on mass starvation in Queensland:

The combination of unusually low winter temperatures and shortage of food has been blamed for the large numbers of flying foxes found dead or dying along much of the East Coast during the last few weeks. With their lower fat reserves and inability to fly greater distances in search of food, juveniles have been particularly hardest hit. . . .

Similarly, a report from inland New South Wales had the gripping headline 'Baby bats abandoned by starving mothers as food shortages kill the flying foxes'. Mothers and babies were fleeing the coastal region, desperate for food because 'storms, deforestation and changing weather has led to less flowering eucalypts and fruit trees this year [2016]'.[33] Reports were that there was starvation all along the coastal region of southern NSW. Flying-foxes had not previously ventured inland at that time of year, and indeed until recently it was not part of their regular range at all. The result was that human carers had no experience in caring for babies, and the youngsters were sent on to Sydney and other coastal centres to be cared for by experienced people and to be released back into the bush. The dramatic experience of realising that something awful was happening was perfectly evident to local people. According to a wildlife carer, 'These bats are in deep trouble as there is a state-wide maternal starving event happening.'

Every year without a new cohort of young ones is a year's loss of reproduction, and with starvation events occurring frequently Greys, in particular, are pushed ever closer to the edge. Louise Saunders was both eloquent and distraught when she spoke of starvation; her participatory involvement with flying-foxes meant that she was extremely attentive to the weather in her area, and elsewhere:

We get starvation events every two–three years here . . . We had one in 2010 that was right up and down the east coast . . . That was very nasty here. We had bats staying in cocos palms until they were eating green fruit. People would ring up and say 'We've got dead bats in our trees, come and get them.' And we'd say, 'Well, if you'd called us a few days before we could have saved them.' People just weren't thinking. They were thinking they were dying of disease, and I'd say 'no, it's starvation'. When people realised, they were horrified that they'd let them die. That was 2010.

In 2011 there was the most amazing flowering here in the south-east. And then another one in 2012 – another starvation event.

The Greys do it really tough, we had a wet period last year. Must have been at the beginning of the year. Must have been ten weeks of solid rain. I got called to a rescue, and it was paperbarks, they were fully in flower. Beautiful. Gorgeous. And there were bats dead underneath. Because there is just no nectar [the rain having washed it out]. The Greys in particular, they're such nectar feeders.

It didn't affect the Blacks so much because they eat leaves and fruits and all sorts of things. But the Greys are so much more [focused on nectar]. And imagine the Reds – they'd just move on. But the Greys stick around. And these cocos palms are giving them false security as well because they're everywhere.[34]

I asked about the cocos palms, and Louise said they're toxic and give the flying-foxes constipation. This introduced plant has spread so rapidly that it is now regarded as a weed of national significance. A tall palm that seems to beckon towards tropical holiday pleasures, it is a favourite ornamental in both home gardens and public areas, and it is detrimental to flying-foxes in a number of ways including those Louise spoke of. Their feet can become entrapped in the branches. In times of hunger, flying-foxes may fly down to the ground for fallen seeds; there they are in danger from dogs and cats as well as cars.[35] Wing membranes can be injured by the fronds, and while membranes can self-repair if the rents are small, large tears make it impossible for them to mend and thus they are unable to fly.

Pests and profits

What is a pest? There is no technical definition. In everyday language as well as in the language of experts on pest eradication, a pest is a type of creature that gets in the way of human projects or human comfort.[36] Pests annoy people, damage property and interfere with both individual livelihoods and corporate commerce. The most prevalent approach to pests is death. In Australia, there is considerable public attention on pests. Diverse resources, both public and private, are devoted to the extermination of species in conflict with human comfort and wealth. A Cooperative Research Centre (CRC) is a government initiative, supported by government funding, which links industry, university research, stakeholders and others for the purpose of developing practical knowledge and capacity on a specific topic. The Invasive Animals CRC focuses on vertebrates that are deemed to be invasive, either because they are not native (they are introduced), or they have 'gone bush' (ferals), or because they have become identified as a 'pest' for other reasons. One species can fit in all three categories, of course. The purpose of this organisation: 'The Invasive Animals CRC creates new technologies and integrated strategies to reduce the impact of invasive animals on Australia's economy, environment, and people.' Technologies, when we explore the term, turn out to be focused primarily on killing.[37]

The CRC's definition of 'pest' is that it is 'an animal that causes serious damage to a valued resource. Such a pest may be destructive, a nuisance, noisy or simply not wanted.'[38] This open-ended definition prioritises human views on what constitutes a nuisance or is unwanted or destructive. On the basis of variable and idiosyncratic views, decisions are made about what is a pest and thus what is legitimately to be targeted with technologies of death. In educational materials linked to the CRC, flying-foxes are listed as pests in a category of native animals that also includes magpies and possums. I am vividly reminded of Donna Haraway's insight that the stories we tell are *of* the world, not only *in* or *about* the world, and so they have life and death consequences.[39] 'Pest' is just this kind of story. When flying-foxes are classed as a pest, rather than as a keystone species or a threatened species (both accurate and both precisely defined), they are slotted into an existing category that is negative and that already has a solution: death.

Shooting was and remains a primary technology in the battle against flying-foxes. A photo from early 1900, taken near the Gordon Camp in Sydney, shows a group of about forty men posing for the camera. They are neatly dressed in dark pants and white shirts. Some wear ties, some wear hats and most are holding a rifle. A few individuals appear to be children. According to the accompanying text, 'In the mid-1800s to the early 1900s it was not uncommon to spend a day out on the weekend on a shooting expedition. Mothers would pack a picnic and fathers would teach their sons how to use a rifle.' This group was planning to shoot flying-foxes.[40] By the early twentieth century the war against flying-foxes was becoming more of an organised effort. '"In the first three years of its existence the Brisbane and East Moreton Pests Destruction board accounted for nearly 300,000 flying-foxes" under a bounty system and probably at the rate of threepence a head!'[41] More recently, there were estimates of 100,000 or more grey-headed flying-foxes being shot annually in the 1990s.[42] The slaughter is contentious, and may have abated, but it is not a thing of the past.

The idea that we humans would righteously blast away at something that annoys us is not comfortable for most people these days. A lot of tricky work is accomplished with language: words such as management, abatement and similar open-ended terms sound officially neutral and are used to cover and perhaps normalise a range of cruelties. I use the word 'kill' rather than 'cull' in order to name deathwork with greater precision. Martin and McIlwee offered the term Pteropucide to name the man-made mass death inflicted upon flying-foxes. Their focus was on the shooting undertaken by orchardists who were seeking to protect their fruit crops. Many orchardists have or had a grievance against flying-foxes. Although the evidence is clear that flying-foxes, Greys in particular, prefer the

Myrtaceae plants with whose blossoming and nectar they are co-evolved, the clearing of native vegetation and its replacement with commercial fruit crops has left them little choice but to go for the fruit. Evidence shows that the level of crop damage correlates with extinguished native foods.[43] Shooting was and for some may still be a first-line response.

Advocates for flying-foxes, including the indomitable Carol Booth, have investigated the effects of shooting on flying-foxes. Booth and others are, of course, mindful of the effects of mass death on the future of the species, but they also offer the necessary per-spective of the individual flying-fox's experience. Legislation con-cerning animal welfare is an important part of the story. Booth was the lead author on the powerful report, *Why NSW Should Ban the Shooting of Flying Foxes*. One of the reasons is that it is inhumane. A study of 155 flying-foxes which had been shot in NSW orchards (dead or wounded), 'found that only 5% had been shot in the head. Autopsies of 30 euthanased flying-foxes revealed "severe injuries, including multiple compound fractures to bones ... that led to incapacitation but not death".' Deaths of lactating females would result in death by starvation for their young. The analysis showed that 40 per cent of the flying-foxes were lactating, as the fruit season coincides with the breeding season.[44] Similarly, a recent (2017) raid on a property north of Brisbane, organised by the RSPCA, found fifty or sixty dead or dying flying-foxes. 'The RSPCA said the bats were found with their wings shot off, bodies riddled with pellets and babies clinging to their dead mothers.' The report offered the bland reassurance that, 'animal cruelty charges were possible'.[45]

The authors of the Invasive Animals CRC educational material state that one way to achieve a successful eradication is to ensure that 'the control operation can remove pests faster than they can reproduce'.[46] On this criterion, Greys are clearly on the way towards eradication. Martin and McIlwee take the analysis further, showing that shooting, while intolerable because of the cruelty it inflicts, and its impacts on a threatened species, is also part of larger pat-terns of lethality. They use the metaphor of a black hole to analyse the dynamics of orchard death. Their scathing study offers the solution to the puzzle of two conflicting ideas about flying-foxes. Orchardists claim that the number of flying-foxes is increasing. They point to the fact that they kill, and kill, and still there are more flying-foxes. Clearly, from their perspective, killing is having little or no impact. Scientists, on the other hand, are certain that the numbers are decreasing. The solution to the puzzle is that orchards, or any other places that offer food, draw flying-foxes towards the nutrition they need. The 'pteropucidal black hole' dynamic is this: every place which affords food is an attractor, and if the local population has been killed or driven away, more creatures are attracted. 'The culling produces a local vacant niche, which becomes occupied by animals moving into it from further afield, which are then killed, so producing a local vacant niche which . . . and so on.' They refer to these kill/attract/kill zones as pteropucidal, and they attest that the dynamic is like 'an irresistible gravitational force sweeping everything into its maw'. The inexorable dynamic works with the forces that drive and draw flying-foxes to orchards, and it creates zones of attraction which become zones of injury, suffering and death. The ecological term is 'source-sink dispersal'; the orchard is a sink (in the sense of sinkhole) into which more and more flying-foxes enter and perish. For a 'grower . . . who is killing the flying-foxes, there will be a perception of "millions" of animals – a never-ending supply – and the misconception that the animals breed like rats and mice.'[47] To advocates and officials who are committed to protecting vulnerable species, the black hole is a biocultural matrix of amplifying death.

The state of NSW has recognised the problem. Flying-fox advocates were brilliant in their efforts to ban shooting in this state, but it is still legal. What was achieved eventually was a commitment to extensive, high-quality netting. There is an excellent and readily accessible literature describing the process in immense detail.[48] Tim Pearson's vigorous words tell the story:

The really interesting thing about that is that enough pressure was brought on the State Government that they actually commissioned an independent panel to look at the issue. Well, they weren't going to release the results to the public until they got pushed with an FOI request, and then they published it with minimal fanfare . . . I can't remember, off the top of my head, the exact composition of the panel. They came to the conclusion that the practice of shooting flying-foxes for crop protection was unethical, immoral, almost definitely illegal, and if it hadn't been for the convenient fact that in New South Wales legislation only the RSPCA can bring actions for animal cruelty, would lay both the orchardists and the government open to legal action. They came to the conclusion that it was also ineffective . . . and that the only reasonable method of crop defence was exclusion netting.[49]

If orchards are well netted, flying-foxes can't get at the fruit. The NSW Department of Primary Industries and the Rural Assistance Authority developed a scheme to help orchardists by providing fully secured netting.[50] Public money has gone into the netting, as is proper. It is in everyone's interest that flying-foxes not be persecuted. The rate of uptake is good. The results improve the harvest, and not only because flying-foxes have been excluded. Nets exclude birds, rats and possums, protect fruit from wind and hail, and produce a microclimate that is beneficial to the fruit.[51] Benefits translate into quality fruit that may set earlier and fetch higher prices. In addition, less time and effort is expended shooting and disposing of flying-foxes.[52] According to Tim:

Typically, [in the] Sydney basin and the Central Coast, somewhere between . . . depending on who you talk to, 75 per cent and 85 per cent of orchards in New South Wales are netted, simply because most of the commercial orchardists who are in areas with flying foxes worked out a long time ago that nothing really worked as a deterrent, so all you could do is put up netting, and although it costs a bit upfront, it was actually worth it. In fact, some of them can make the money back – in a good year, they'll make the money back in one year. In a series of bad years, it takes them five years to recover the cost of the netting. It's the small, largely family-run orchards in the Central Coast and Sydney Basin who don't want the net. They claim they don't have the money for it, especially in the Sydney Basin up north. A lot of them are traditionalists.[53]

Like any group of people, orchardists are diverse. Carol Booth came face to face with baby flying-foxes through friends she made in northern NSW:

I lived in Byron Bay in the early nineties. I saw someone with a little baby flying-fox and found out who the carers were. They were a really interesting family. They were fruit farmers, and they netted their crops and became flying-fox carers. Such lovely people. I went and met them and got to care for a baby Black flying-fox. Gorgeous creatures! Things were so cavalier back then. No regulations. It was really such a fun, unstressful experience. I just fell in love with them.[54]

Still, it must be tough. I imagine flying-foxes circling the edges of orchards knowing food is there but not knowing how to get it. Perhaps they are wondering where they will go to feed themselves, night after night. Starvation or shotguns! I am projecting an idea of choice that flying-foxes almost certainly do not perceive, but the problem is real. One scheme to alleviate this dilemma is to ensure that tree planting programmes now in process – for timber, or as part of Landcare initiatives, for example – focus on species most needed by native animals. For flying-foxes this would mean planting trees that flower in winter and spring.[55]

Advocacy and electrocution

One of the great advocacy stories takes us into the legal sphere of animal protection and has had some positive and significant outcomes; it involves lychee farmers in Queensland's far north, and the flying-fox activist and advocate Carol Booth. Carol's calm demeanour rests lightly on the passion of her commitments. Having listened to her in person and read her well-reasoned and professionally phrased written work, it was a shock to see news clips showing her being dreadfully harassed. Carer-advocates admire her greatly for her strength of commitment, gritty determination, and canny ability to negotiate bureaucratic and legal contexts. Nick Edard, the man who was so active in trying to stop the Sydney Royal Botanic Garden expulsion, said: 'One of the other issues around flying-foxes in New South Wales is shooting. One of the people we admire very greatly is a lady called Carol Booth, who was instrumental in getting shooting struck off in Queensland.'[56] As we have seen, however, permission to shoot was reinstated a few years later.

The court cases brought an experienced advocate together with science professionals, and several government and non-government agencies and organisations. The cases I discuss here concern the use of electrocution grids to kill flying-foxes that came to feed at tropical fruit orchards. These grids had been legally banned in 2001 on the grounds that they were inhumane. Subsequent legal action, then, concerned enforcing that legislation. There is no knowing how many flying-foxes were killed on the grids, but through surveys, and from farmers' own evidence, the figures were high. In *Booth* v. *Bosworth*, 'the court found that an estimated 18,000 spectacled flying-foxes were electrocuted during the 2000 lychee season and that continued operation of the grid would cause the species to become endangered within five years'.[57]

Carol had written many of the submissions that argued against electrocution and other forms of cruelty. She had coordinated and co-authored major reports on protection of flying-foxes from cruelty.[58] She describes how she became engaged in this work:

I grew up in the Bjelke-Petersen era in Dalby [Queensland], a very conservative, agricultural area. If you love nature it's a bit depressing, really. And I really only became involved in NGOs after I went overseas and lived in China for three years. I think it was that contrast between Australian and Chinese society, where in China your capacity to change things was extremely limited. I came back with the sense that I had more capacity to influence than I'd realised. So when I came back I got involved in local environment groups. When I went to North Queensland I was coordinator of the North Queensland Conservation Council [NQCC] for a while and flying-fox issues became prominent.[59]

In the electrocution cases, she was seeking to require state governments to respect and enforce their own legislation. The use of electric grids to protect orchards was rare in NSW, but in Queensland a large number of farmers of lychee and other tropical fruits sought to protect their crops by erecting grids of wire with electric current running through them. The intention, and the effect, was that as flying-foxes came to feed they collided with the wires and suffered electrical burns. In subsequent court cases, farmers claimed that death was immediate and painless. Research by scientific experts refuted the ideology of a quick and easy death.[60]

Len Martin served as an expert witness in several court cases. Along with his work on demography and the pteropucidal vortex, his expertise included a deep professional knowledge of the capacity of animals to experience pain, so he was an ideal witness to assess the impacts of electrocution grids. He was asked to provide professional assessments of several issues. On the question of demography, he concluded that the grids had the potential to amplify population loss in species already in decline; his main focus was on the Spectacled flying-foxes of the wet tropics. On the matter of a quick and painless death, he completely disproved the idea that death was necessarily quick, and showed that if not instantaneous, it would subject individuals to severe, indeed excruciating, pain that would torment the creature during whatever remained of its life.

Interestingly, in NSW two farmers using electrocution grids had been brought to law on charges of animal cruelty, as it was defined in their state. The case was that the grids caused 'multiple uncontrollable acts of cruelty'. In due course the farmers were fined, and the conviction was recorded. Martin had appeared in this case, so he was surprised to be invited to contribute to cases in Queensland where the farmers were mounting strong opposition to the threat to their grids. Their stated view was that the loss of the grids threatened their livelihoods, and that they would not be dictated to on matters concerning their own property. In spite of the earlier legislation outlawing the grids, angry farmers claimed that death was instantaneous, and thus that no suffering was involved.

Martin drew on previous work on flying-fox demography to conclude that the impacts of the grids would be significant.[61] His work on the linked issues of pain, suffering and cruelty was both insightful and alarming. He drew on a range of research conducted by numerous scientists to conclude that:

> a large body of objectively verifiable knowledge supports the view that all mammals perceive pain in essentially the same way . . . It also seems that for 'pain' to be of adaptive value it is essential that it be a consciously perceived sensation . . . Thus the accepted scientific view is that all mammals perceive pain in essentially the same way and to the same degree as humans.[62]

The idea that death was instantaneous was shown to be false by a survey of dead flying-foxes in a grid-protected orchard conducted by Carol Booth and Allen McIlwee. They found living animals still suspended on the wires, and on the ground they found more injured but still living creatures, including infants who had survived the mother's encounter with electric wires. In non-lethal encounters, the effects of electricity cause long-term suffering. The electricity itself burns creatures. Flying-foxes most frequently encounter wires when their wings, along with the hands and feet, become entangled. The evidence from powerlines is a consistent guide to the effects of grids: 'deep burns, even to the bone, on the forearms, thumbs and hind limbs (sometimes the hind claws are fused together); large areas of wing-membranes burnt away with much of that remaining crisped, brittle and crumbling'.[63] Individuals injured in these terrible ways may yet live for hours or days; if found, they would almost invariably be euthanased. Martin writes that 'many flying-foxes, making the[se] electrical contacts, will be conscious and in severe pain, and either struggling, or undergoing reflex or electrically induced spastic movements for some time before they die. Such movement and spasms may throw them from the wires severely injured and in pain but alive.'[64]

Martin's conclusions addressed the fundamental question put to him: did the grids inflict multiple uncontrolled acts of cruelty? His answer was yes. Multiple acts were inflicted as evidenced by the fact that so many flying-foxes were affected. The impacts were uncontrolled because there was no way to limit the numbers (or to control for sex, age and maternal responsibilities) of the flying-foxes. And the acts were cruel in that:

> as a consequence, animals are unreasonably and unjustifiably mutilated, maimed, terrified, exposed to excessive (electrical) heat and inflicted with pain. Similarly, that operation of a grid results in multiple uncontrolled acts of aggravated cruelty, in that its operation will result in the death or serious disablement of multiple animals, some being so severely injured or in such a condition that it is cruel to keep them alive.

He recommended that all such grids be dismantled immediately, but he kept a final word for the wider public because there still remained the problem for farmers of how to protect their crops: 'if the community wishes flying-foxes to be conserved, then it must help fruit growers with non-means-tested grants and subsidies for humane, non-lethal methods of protecting fruit crops'.[65]

When Carol Booth first learned of the electrocution grids she was moved to action, and with the support of other carer-advocates and through her work with NGOs she took on the issue through every available legal opening. One lychee farmer in particular, Rohan

Bosworth, was vigorously committed to his grids and incredbly angry about being told what he could and could not do on his property; a number of the court cases involved *Booth* v. *Bosworth*. There were appeals, and more appeals; the case went back and forth and was heard by several judges. The net result was that the state was persuaded to enforce its own legislation.

Carol's words again:

> Well, the first one was the Bosworth case. So that was initiated under the federal Environment Protection and Biodiversity Conservation Act . . . There was this real property rights flavour to the opposition, just the shock that these new environment laws required farmers to change their habits. They really couldn't quite believe that it was possible. . . .[66]

Legal action for protection was directed towards two possible lines of argument: the status of the flying-foxes under federal laws, and questions of human cruelty and the suffering experienced by flying-foxes.

On the matter of the legal standing of Spectacled flying-foxes as a species, the legal rulings were significant. The species is not listed internationally as being threatened, but its key role in sustaining the wet tropic rainforests was recognised, and the species was granted protection for its keystone role.[67] Justice Branson held that the loss of a single species could constitute a significant impact: 'This decline would undermine the capacity of the spectacled flying-fox to contribute to the genetic and biological diversity of the World Heritage Area.'[68] The last of the cases Carol Booth brought to court in 2006 with the backing of the Environmental Defender's Office shows the intransigence of some of these farmers. According to the EDO's summary: the case was against farmers named Yardley who had continued to operate their electrocution grid. Booth sought 'to restrain their use of the electric grids, to require dismantlement of the grids, and to require a financial contribution by the Yardleys for the rehabilitation of flying-foxes'. The judge ordered that the grids be dismantled, and: 'When the Yardleys failed to dismantle the grids, our client (Booth) brought proceedings for contempt, in 2007 and again in 2008. After the second contempt proceedings were initiated, the grids were dismantled. The Yardleys were fined $5000.'[69]

From an analytic point of view, the shifting status of Spectacled flying-foxes and the recognition of their keystone standing in relation to the tropical rainforests was an outstanding achievement. It also highlighted diversity and conflict around questions of property. Farmers defended their right to kill because they believed they should not be told how to manage their properties. On the other hand, Judge Branson identified the World Heritage rainforests as another type of property – heritage property, responsibility for which comes from both the Australian government and international agreements, and which is exercised on behalf of the nation and the world. Two types: one local, private and privileged in a long legal tradition of respect for private property; the other international, collective and privileged in the domain of universal values and action. Looking at Judge Branson's decision through an ecological lens, we see a decision that implicitly asserts the value of connectivity, symbiotic mutualism and ripples of effects both positive and negative. Still couched in legal language, it was and remains a significant achievement.

As I read the summaries of the court cases, though, I am struck by the small amount of attention given to questions of cruelty and suffering. These issues had already been assessed and had resulted in the ban on grids, but they remain at the heart of all further legal action. One author wondered if the evidence was just too awful for people to want to linger on. And yet, it is the baseline of the ethical and empathetic response people experienced when even thinking about, much less confronting evidence of, all that horrific misery. The legal framework made the case more manageable from the perspective of human emotions, and was focused, properly, on what needed to be achieved to ensure enforcement of and compliance with the law. One cannot read the summaries, though, without feeling a certain wounding of one's own conscience. And so it remains the responsibility for all of us who work outside the courtroom to bear witness to the fact that we are in the midst of ongoing, soul-affecting deathwork.

Spillovers

The metaphor of gravitational draw is powerful in itself, and can be taken further: the rapacious maw of the pteropucidal black hole does not have a boundary that stops with flying-foxes. As we have seen, when flying-foxes are dragged into the vortex of death, the long-term ecological effects mean that open forests, rainforests and other ecosystems are dragged along with them. From the point of view of forests, it is some comfort to know that flying-foxes are not the only pollinators, but deathwork is widespread. Birds and insects are also in decline and many are in serious trouble.[70] This means that critically endangered ecosystems are being dragged into the vortex, and so will the rare and endangered plants and animals that live in them. As is well known, rainforests are colloquially referred to as the 'lungs' of the planet, soaking up carbon dioxide and pumping out oxygen.[71] As rainforests disappear, so does the possibility of sustaining an earth system that will be inhabitable for large numbers of the species of living beings who have evolved here and belong here.

The black hole does not exempt humans, and this is so in immediate ways as well as in the longer-term prospect of losing Earth's habitable climate and atmosphere. One of the most alarming new factors in expanding ecological destruction is the emergence of new viruses, many of which spill over from animals. The term is zoonotic, meaning that animals are the source, and Chiropterans (flying-foxes and bats) are prominent carriers of newly identified viruses.[72] The three diseases relevant to Australian flying-foxes are Hendra virus, Menangle virus and lyssavirus. Scientists who work in this field tell us that these diseases are not in fact 'new' although they seem so because they were all identified in the 1990s. They are actually of ancient lineage, and apparently have only recently been 'dislodged from their normal niche as a result of (generally anthropogenic) changes'.[73] Epidemiologist Hume Field explains:

> It might be that wildlife populations can cope and cope and cope with impacts, until they get to a certain threshold where their ecology is fundamentally compromised . . . And it's those points . . . those tipping points, that can precipitate the emergence of a new disease.[74]

Hendra virus affects horses and is transmitted from horses to humans. There have been a number of deaths of both horses and humans.[75] The method of transmission is uncertain but probably involves flying-foxes defecating into water troughs from which horses drink. Since 2012 there has been a vaccination for horses, and a public awareness campaign teaching people how to protect themselves when handling sick horses seems to have put an end to human deaths.

Menangle virus afflicts pigs, and it seems probable that humans get it from pigs. There are no reported deaths.[76]

Australian bat lyssavirus (ABL) is closely related to rabies. The vaccination is completely effective but must be kept up to date. The disease is transmitted via a bite or scratch and is fatal if not treated quickly; it can be prevented, but must be treated immediately with injections of human rabies immunoglobulin (HRIG), followed up by vaccine injections. Only three people are known to have died of ABL, one of them a little boy who was scratched or bitten and died in 2013. The disease is transmitted directly from flying-foxes as well as from other bats, including microbats. Everyone who handles flying-foxes is advised to be vaccinated, and in numerous contexts people are required to do so. Given that the disease is fatal, precautionary measures are always appropriate. When I was bitten by a flying-fox I was strongly advised to get more rabies shots even though I was already well vaccinated because the training course I took required proof of vaccination.

One of the most interesting questions is why the flying-foxes themselves seem to be largely immune to the lethal viruses they carry. The answer is in their immune system, but that is just a first step. Research is ongoing into how Chiropteran super-immunity works, and how it differs from the human immune system. Perhaps there is a prospect for finding ways to boost human immunity to many illnesses.[77]

Because these diseases are actually ancient, the question arises as to why they have suddenly appeared. The answer involves several interconnected factors. With urbanisation, flying-foxes and humans spend more time in proximity, and there are more opportunities for contact. Equally, flying-foxes are becoming more sedentary and there is evidence showing that their migratory way of life enabled conditions that contained the virus within the host. A more general explanation includes the first two but goes on to suggest that stress is an important factor in weakening flying-fox immune systems. An important study of Hendra virus showed that spillovers correlate with the birthing season. Factors in the flying-fox life cycles were part of the varying strength of the immune system, and mass birthings were driving disease incidence.[78]

Flying-foxes are subject to many stresses that affect their immune systems, as we have seen: harassment, dispersal, death of family members, starvation, wounding from guns and electrical wires and other hazards such as barbed wire, to recapitulate a few. With these emerging diseases harm comes full circle and demonstrates the inextricable connectivities between human health, flying-fox health and habitat health. The vortex keeps growing. More stress means more spillovers, more spillovers may lead to more fear and more demand for actions such as dispersals that generate enormous stress. The language of vilification of flying-foxes gains new ammunition. As flying-foxes and humans move closer and more permanently into contact with each other, so new viruses threaten both humans and flying-foxes, expanding and amplifying the dynamics of death.

The best response, as Carol Booth and others point out, is to 'conserve flying-foxes and reduce the environmental stresses –

including shooting – that increase their rate of infection and the risk of spillover to other species'.[79] If, however, the human response is to accelerate the stresses in an effort to control the boundary between humans and flying-foxes, the biocultural loop takes on the shape and energy of a vortex. Devastation amplifies: disastrous, entangled and recursive.

8 Cruelty and its Allies

A British biologist named Francis Ratcliffe was invited to Australia in 1929, sponsored by the state governments of New South Wales and Queensland, to investigate the orchardists' problem. He was asked to provide information on flying-foxes with the aim of determining how best to get rid of them. It became clear to him that total eradication of flying-foxes was probably not achievable in the short term. He offered a consolation, though: the populations seemed to be in fairly rapid decline, and thus it seemed probable that the 'problem' would take care of itself.[1] His further insight was that the problem of this 'economic pest' was greatly exaggerated: 'a mythical idea of the appalling destructiveness of the flying fox has gained ground. It completely dominates all popular writing. . . .'[2] The exaggeration he perceived then continues today, only now urbanisation triggers demands for dispersal.

Warfare

Ratcliffe remained in Australia, and his book about his experiences in arid regions 'achieved the status of a minor classic'.[3] His technical report, published by what is now the CSIRO, is a model of thorough research. He identified, evaluated and offered suggestions for the improvement of a large number of lethal methods taken by orchardists and by 'professional flying fox killers'.[4] He referred to the animals as either 'pests' or 'foxes' throughout the evaluation. The criterion for assessment was the cost-benefit ratio, that is, cost of killing in relation to the value of the fruit harvest. Thus, for example, his negative evaluation of poison gas was based both on its limited lethality and on the cost of obtaining the gas and effectively delivering it to the target. Ongoing hazards for humans were mentioned in one case. The assessment was extensive, and it is worth examining because it provides a well-researched snapshot that illuminates violence in language that is not self-censored, as far as we know. The idea that a society would seek to exterminate a native species seemed to be completely normal.

- Camp destruction – would work best if major sites to which flying-foxes returned regularly were completely destroyed – could work well if there were affordable methods to completely destroy a large area of trees (59)
- Biological control/Introduced diseases – almost no chance of success because of the 'difficulty of initiating an artificial epidemic' (64)
- Scalp bounty system – seems practical according to popular thought: 'If an animal is an economic pest it should be killed, and if a sufficient bounty is put on its head, it will be killed.' However, this method costs more than the damage flying-foxes do; there is ample scope for fraud; this is the least effective method (65–7)
- Commercial use of skins – ineffective. Could overcome the problem that the bounty system costs too much, but the pelts were determined to be 'totally unsuited for commercial purposes' (68)
- Poison gases (chlorine, hydrogen cyanide) in camps – ineffective. Only chlorine is really cheap. The gas is heavy and most of it remains low to ground. That which rises irritates the flying-foxes and they leave. Gas may linger near ground level and pose a threat to humans (69–70)
- Flame guns – costly, but promising under the right conditions.

According to a description of flame-thrower impact: 'dozens fell down burnt to cinders and dozens were so badly burnt that their death was only a question of time'. Might be rejected on grounds of cruelty (70–1)

- Explosives – 'complete failures': the trees were damaged and the flying-foxes took off. Could be useful if the bombs were fitted out with shrapnel (71)
- Shooting – semi-organised or recreational 'battues' largely carried out for sport have no consistent effect. If organised and pursued over a number of years, professional shooting might have 'beneficial effect', particularly if 'females carrying young' were targeted (72)
- Poison baits/Strychnine – 'well worth the trouble' in certain orchards (74)
- Deterrents – for example, stringing barbed wire across the orchard so that flying-foxes become tangled and eventually die: too expensive. Other deterrents either ineffective or effects uncertain (75)

Ratcliffe used the language of warfare to express his mission:

> When a military commander plans an offensive he must have certain information on which to work. He needs to know, for instance, the size and strength of the enemy forces and the tactics they are likely to employ. In the same way, before a campaign is launched against an animal enemy, it is essential to have accurate knowledge of its numbers and habits. . . .[5]

Some of Ratcliffe's prose can perhaps be read as if he were indulging in tongue-in-cheek humour. I am not sure. His cost-benefit assessments of deathwork show an economic-managerial paradigm that completely excludes the living beings whose lives are at stake. The suggestion that shooters target females with their young clearly indicates that Ratcliffe understood extermination as an intergenerational process, and he seemed to be unconcerned by any feeling for the animals whose deaths he was helping to orchestrate. There was no evident concern for animal welfare other than to note that some people might object to the use of flame-throwers. It might be thought therefore that in those decades people did not object to animal cruelty, but that would be wrong. Concerns about cruelty have a long history in the West, going back in recent centuries at least to the essays of Montaigne; they are addressed philosophically and are ethically grounded in the observable fact that animals suffer. The pre-eminent philosopher of the question of suffering was Jeremy Bentham (1748–1842), and his outstanding question is still relevant: 'Why should the law refuse its protection to any sensitive being?'[6] Activism gained social momentum in the Victorian era with the anti-vivisection movement, and the commitment to animal welfare continues. We have legislation concerning animal welfare, there is a field of animal law, and there are animal-oriented organisations such as Voiceless whose members advocate on behalf of those whose voices are often not heard.[7] Let's remember that many of the stories told in the public sphere have shadows which obscure immeasurable violence. Flying-foxes and many others need advocates. Voicelessness is not the same as silence, for animals too participate in the languages of the world, but in the public sphere where human voices dominate, nonhumans need advocates.

Language is central to violence, and it is important to analyse both language and action. The goal of staying with the trouble also includes generating some counter-trouble. I want trouble that cuts across a public culture that tries to normalise suffering, cruelty and violent death, and that uses language as a protective shield. I will, therefore, enter more deeply into zones of persecution. I am something of a reluctant witness in these matters, most particularly because I dislike giving hatred, violence and persecution any more publicity than they already have. And yet, much of the public/political discourse that concerns the persecution of nonhumans must be disrupted.

We are in the midst of attacks on Earth's shimmer.

* * *

To return to Ratcliffe and his language of war, remember that warfare implies combat conducted according to rules. There are categories: 'just war' or 'war crime'. Warfare involves soldiers killing soldiers, and winning is substantiated through their injury and death.[8] Success is achieved when one party surrenders, and in twentieth-century warfare there follows a period of treaty-making and post-war reconstruction. Clearly, the battle against flying-foxes and other animals is nothing like this. It has never involved equals, and flying-foxes are not, themselves, engaged in warfare. They are just living their lives as best they can under increasingly difficult circumstances. To be fair, contemporary warfare is not much like this ideal model, either; better language is needed.

Knowing the enemy is one side of Ratcliffe's vision of war. Another side, which he apparently takes for granted, is that of building up public enthusiasm for war. This has meant extolling the righteousness of the cause, while vilifying, perhaps demonising, the enemy. Enthusiasm and hatred are linked so as to articulate a public culture that willingly commits to violence. Truth has no particular place in this discourse. Mostly we encounter the language of wild exaggeration. One has only to think of the horrific Nazi iconography of Jews, and the American vilification, and incarceration, of the Japanese. A remarkably uniform vocabulary of vilification/falsification emerges in both language and imagery.

In Australia, when an animal is declared a pest, death becomes its destiny. Suddenly, whatever it does is wrong in the eyes of those who are determined to get rid of it. And suddenly wherever it is, that is where it must not be. With flying-foxes, some of the groundwork for hatred and fear has a long history. When Pteropids are equated with microbats, as they often are, then a whole complex of issues can be enlisted: not only are they nocturnal, and have a particular wing shape, but in the more fanciful domains of the human imagination they are vampires, perhaps harbingers of evil, perhaps even devils themselves. And so they can be evoked in public imagination as super-natural enemies as well as natural ones.[9]

Anyone who has ever listened to shock-jock talk radio has a good idea of what vilification sounds like. The aim is to generate fear, hatred and outrage through language and tone of voice, presenting emotional responses as if they were inevitable facts. There is right-eousness, outrage, a flow of denigratory words; outrage is amplified by connecting disparate issues also deemed to be outrageous, so a story of conspiracy appears to emerge.[10]

An example involving a Sydney shock-jock was triggered when a group of flying-foxes left a site where they weren't bothering anyone; rather, they were being bothered by the construction of a desalination plant near their camp. They left one night, and they took up residence at a site adjacent to a school for special needs children. In the course of one relatively short rave, the host spouted off a good selection of vilifying lies: 'Bloody flying-foxes, an absolute army of them. These things are vermin, they're pests' . . . 'Ferals' . . . 'Three people have been killed!' . . . and there's 'no treatment for bat lyssavirus'. And: 'Listen to this . . . The noise, the health risk, the putrid smell of ten thousand bats is plonked on the doorstep of "this special school".' 'The school was here first!'

The wrap-up hinted at conspiracy to cover up the numbers, and a demand for a 'cull', in spite of the fact they are protected species. The ideas purveyed in this rave were so off the mark that Storm Stanford, one of the supremely committed advocates, issued rebut-tals to eight or so lies, particularly stressing that lyssavirus can be prevented if appropriate steps are taken.

As one would imagine, the flying-foxes weren't allowed to stay at the site near the school. They were dispersed, and no one knew just where they went. Tim Pearson commented: 'Like the Sydney Royal Botanic Gardens, they accept that this will be an ongoing action – probably for some years. The approved budget for this is $800,000 to $1.2 million.'[11]

Extinction debts

These days the main animal villains publicly accused of disturbing our imagined ecological destiny are 'feral' (bush) cats. An attention-grabbing headline appeared in *The Sydney Morning Herald* in 2017: 'War on feral cats: Australia aims to cull 2 million'.[12] The article went on to say: 'The federal government will unleash every weapon in its arsenal to wipe out 2 million feral cats – about a third of the population – and will provide $5 million to community groups to serve as foot soldiers in the battle.' The 2 million figure was just a target for 2020. The programme is set to run for twenty years. Why? The reason for this violence is that bush cats prey on native birds and small mammals, many of which are threatened with extinction. The social commitment to protecting native species is excellent, but the factors leading towards their demise are complex, and actually loss of habitat due to land clearing is identified in scientific studies as the key process.[13] Cats are being made to bear the burden of blame for problems that have human causes. The cultural process of blaming cats is known as scapegoating. It justifies, or seeks to justify, methods of killing that, in this case, include shooting, trapping and a reputedly 'humane' poison. Research is currently underway to develop new methods.[14]

The work of killing cats is, in stated intention, benign. 'We have got to make choices to save animals that we love, and who define us as a nation like the bilby, the warru (Black-footed rock-wallaby) and the night parrot', according to the Threatened Species Commissioner Gregory Andrews.[15] This densely suggestive sentence tells us there is a 'them', and we have to make a choice against them because they are invasive pests. There is an 'us', defined by being members of the nation; we are now people who love our nation both socially and ecologically. Native animals are enlisted in the category of 'us' because they define 'us'. Once they were shot with impunity, indeed often with bounties, as settlers sought to transform Australian ecosystems; now they are important cultural resources for national identity.[16] I can't help but think of Samuel Johnson's great adage that patriotism is the last refuge of the scoundrel. When humans scapegoat cats the appearance is of righteousness, and it lets humans off the hook in relation to politically difficult policies like land clearing. The programme is marked by stated concerns for animal welfare. The RSPCA supports the humane euthanisation of feral cats, and Mr Andrews, criticised by cat lovers as well as lauded by many conservationists, asserted that he had no problem with all this death: 'I sleep well because – having been a cat owner for most of my life – the science says every feral cat will kill three to 20 native animals a week.'[17]

The questions are large and important, ethically difficult, and full of contradictions; they raise again and again the public process of making some animals killable while others are to be protected. There is no unanimity in regard to the damage cats do to native species. The authors of a recent essay on bush cats bring their scientific expertise to the analysis and conclude, on the basis of sound evidence, that both the science and the ethics are shaky.[18] Importantly, a recent report co-authored by the Australian Conservation Foundation, Birdlife Australia, and Environmental Justice Australia exposed the fact that 'successive governments have avoided their responsibility to protect threatened species habitat and have instead entrenched the process of extinction'. The result is our 'extinction debt': a situation established through profound neglect. This debt has left 'thousands of species of plants and animals on a pathway to extinction because of the threats already unleashed and because the area of habitat that has been left for them is insufficient to support viable populations into the future'. The authors make the important point that while governments are shirking their responsibilities, the situation for many animals is by no means irreversible. Actually, 'extinction is far from inevitable for the vast majority of threatened species in Australia. Extinction is the result of the decisions made by successive governments to ignore their own scientific advisers, and to neglect their obligation under our environmental laws to protect the ongoing evolution of life on the Australian continent.'[19] This is a fundamental point: human-made destruction and negligence both are key factors in the biocultural mass-death extinction disaster. The problems are not caused solely by cats. Feline predation certainly has an impact, and every impact matters, but their deaths will not save native animals. Only massive changes to human behaviour can do that.

And so, as shown in discussion of the term pest and the Invasive Animals CRC, violence against certain animals is part of a wider, bureaucratised, government-funded, university-based, industry-supported, socially legitimated business of expertise in killing. People build careers on inventing new ways to kill, and at the same time economic benefits flow to industries producing poison agents and methods of delivery of death. Once it was flying-foxes who were investigated thanks to government funding, now it is cats. Violence is at times cloaked in the language of research excellence, managerial efficiency, at times in patriotism, even mentioning love, but there can be no doubt that what passes for work to produce a more perfect environment by cleansing it of its bothersome creatures is actually work of terrible cruelty, and the vilification unleashed to promote all the death is itself the other side of management language. And the extinction debt continues to grow.

The ethical questions around decisions of who lives and dies, whose lives are worth protecting and whose can be terminated with justification have no easy answers. Thom van Dooren's response is to call for staying with the complexities of 'invasive species' and the problematic of killing for conservation. That is, if there are no simple answers, there are yet possibilities for care and conservation that do not require death as a first response.

In this context, perhaps the goal for a new ecology will be to learn to value and care for what is here now, in a way that holds onto it, but gently; in a way that acknowledges that any given species or ecosystem, while being immensely valuable and precious in itself, is nonetheless a transitory and changing affair.[20]

Persecution has a history

Persecution, vilification and, at times, pride in violence are part of today's public discourse and public policy. They have a long history and are foundational to what the historian R. I. Moore calls a 'persecuting society'. Moore developed this term through his research into medieval European history. He concluded that around the year 1100 western Europe 'became a persecuting society, and . . . has remained one'.[21] He was very clear: it was not just that persecutions happened, it was that they were deliberate and central to society. In medieval times lepers and heretics were persecuted; later it was witches and freemasons; throughout it all there were eruptions of persecution of Jews, and from time to time 'sodomites' were targeted. Along with these categories of human, the dogs, cats and bats associated with them were subjected to torture and death. Persecution was based on stirring up hatred and fear, but it was not only about rabble rousing. Institutionalised persecution was achieved *'through established governmental, judicial and social institutions'.*[22]

Some historians have objected to Moore's conclusions on the grounds that surely all societies persecute those they deem to be outsiders. That may or may not be true, but Moore's idea was that in western Europe persecution became integral to the actual fabric of society. European societies became modern states through deliberate use of persecution. And yet, the fact that targets of persecution have changed with time tells us that the culture is adaptive. Moore concludes that established 'patterns, procedures and rhetoric of persecution which were established in the twelfth century have given it the power of infinite and indefinite self-generation and self-renewal'.[23]

One of the main paradigms which organised the theory and practice of persecution, then and now, is that of the enemy in proximity. As Moore explains, one of the first steps in defining a target for persecution was to develop a story to account for the classification 'enemy'. The target group killed children, or poisoned the water, or spread disease, or . . . the list goes on. We might add from today's evidence: they fed in otherwise profitable orchards, or (as with cats) they killed native species. The cultural pattern is based on the root metaphor of the human body, and it thus taps into deep, legitimate, existential, extremely personal concerns about one's ongoing life.[24] The primary concept is that each body has an inside and an outside; the boundary is permeable. Things get in as well as getting out, and so the boundary always requires defence. Danger lurks. It could be disease; the diseases could be carried by insects or other 'vermin'. The possibility of incursion is ever present. The human body, the core concept, is a microcosm, its macrocosm counterpoint is society, also conceived as a social body or the body politic. In between micro and macro any number of bounded units can be organised according to the same logic. Family + orchard is a good example: the orchard is incorporated into the human domain to be defended, and threats to the orchard equal threats to the family or the individual. Examples abound. But as long as the boundary defines an 'us' and 'them' divided by danger, hatred and fears of incursion, problems abound.

In the case of the flying-foxes that camped near the special needs school, one argument was that the school had been there first. The implication seemed to be that as newcomers, the flying-foxes should leave. Vilification not only ignores facts, it ignores broader, bothersome truths about history and belonging. Who belongs here locally and ecologically? Why are some native species included in the category 'us', like bilbies and night parrots, while others are excluded? How was it that the school was there before the flying-foxes when Pteropids have lived in Australia for millennia, and are known to have been in the Sydney region at the time of British settlement? Vilification matters because it expresses firm grounding for categories that actually are extremely arbitrary and shifty. A social-cultural commitment to violence holds an image of stability in what is uncertain, unstable and unsustainable.

* * *

There was a garden; there were trees, of course, and there were creatures. There were rules; some things were not permitted. But the rules weren't obeyed. There would have to be an expulsion. It is an ancient story, and a foundational story for narratives of origin in the western world.[25] The contemporary story I refer to here does not ground itself in Biblical antiquity, and probably it would not be wise to do so since someone would have to play the role of G-d and there are no gods in this story. The Garden was Sydney's Royal Botanic Garden, the creatures were flying-foxes and the trees were 'specimens'. The language was all about good management as construed and overseen by the Botanic Garden Trust and as mandated under legislation; governmental responsibility rests with the NSW Office of Environment and Heritage. Responsibility for the welfare and management of threatened species is also within that government department. The trees most affected by flying-foxes were heritage trees, deemed to be valuable because of their rarity; they were non-natives and they felt the impact of the continuous presence of flying-foxes.[26] Greys are a threatened species, and are subject legally to protection. The Royal Botanic Gardens has a statutory duty to protect tree specimens. Surely this was the perfect opportunity to think about symbiotic agreements and mutual respect. And yet, the only option the Garden Trust was prepared to adopt was to expel the flying-foxes.

Sydney's Royal Botanic Garden had been an occasional home for Greys for many years; numbers rose and fell, but from about 1989 there was a permanent presence. Within a few years the Garden had begun efforts to expel them, and it was briefly successful. The fact that the site had become a traditional maternity camp put constraints on the methods used to try to get them to move on. Over the years tactics included noise harassment, and ingenious methods such as lacing trees with bundles of python excrement (olfactory deterrence; pythons are one of their main predators) and with fermented prawn paste (taste aversion). Noise harassment is stressful; the equipment used in the Garden was the Phoenix Wailer, a computer-controlled system that blares out a variety of electronic sounds, randomly selected, to create 'a whirling effect of reverberating noises that creates a "discomfort zone"'.[27] When first trialled in 2001 the effect was to reduce the numbers of flying-foxes in the target area by 90 per cent. Those who refused to leave were males who had staked out mating territories and were completely unwilling to give them up. Many of those who left returned once the harassment ceased.[28] These methods had the effect of getting some flying-foxes to shift, but the fact is that the Garden is a great home in what for flying-foxes is a world of shrinking opportunities. Nick Edard, a fiercely committed advocate, described the main parameters of the Garden's appeal:

> It's a very successful breeding camp. Because [for] the flying-foxes there, a large part of their diet is not rural native food sources, it's street plantings in the eastern suburbs, which are a more regular source of food than some of the native and natural food sources, which can be affected by climate changes. The camp is buffered against heat stress. Even during the worst heat stress conditions that we've experienced in the last ten years, the Botanic Gardens has not suffered a major heat stress event. Cabramatta and some of the other colonies, thousands of flying-foxes have died in a single day.[29] That hasn't happened in the Botanic Gardens. It's right next to a huge body of water. The humidity's always going to be better. You have on-shore breezes. There's a lot of areas in the Gardens that are reasonably shaded for them. So you add all these things together and it is a good roost site for them. If you choose to disrupt the camp, it almost goes without saying that it is going to have effects on them, and part of the problem is that it's difficult to prove what those effects will be. They're very unpredictable. They will almost certainly, some of them, join other camps. It may take months for it to happen, but then they're in competition for the food sources with the animals in those camps.[30]

In 2010 the Botanic Garden was granted permission by the federal Minister for the Environment to embark upon a thirty-year process of expelling the flying-foxes through the use of noise harassment. The procedure was designed to cause pain and distress. That is what it takes to break site fidelity: trauma. Because of that strong fidelity, as well as the attractions of the Garden as a place to camp, the expulsion process must go on year after year. Once approval was given there was a lot of debate and protest, and every major point was made: trees can be netted, it doesn't have to be either-or; a botanic garden should be especially attentive towards a keystone species that is so crucial to trees; causing stress to a threatened species is not appropriate; expelling members of a threatened species from a maternity camp is not appropriate; stressing pregnant females is not appropriate; co-existence is possible, would cost less than expulsion, and would set a benchmark for good practice.

Along with advocates' concerns for the lives of flying-foxes, they anticipated that if the Garden expulsion went ahead it would open the door for approvals elsewhere, most of which would not be under the strict scrutiny that the Garden would receive. Once it became clear that the Garden was 'cutting corners', concerns for the future took on new urgency. I interviewed Nick in 2010 when the advocacy shifted from public relations to legal forms of redress. Nick was happy to talk about his involvement in learning to deal with government agencies. He and Storm Stanford worked together:

> I think it must have been the beginning of 2007 – there started to be some media around the dispersal of the flying-foxes from the Botanic Gardens, and a couple of us . . . we weren't real happy with the unopposed media the Botanic Gardens Trust were getting, and we didn't think that what they were telling the media was the whole story. It was a very one-eyed view of events, and we thought that they were really being a little bit sneaky in trying to get public opinion behind them, and probably as an unwanted side effect of that, it was an element of demonising the bats: the bats are a bad thing, they're destroying the trees, this is a heritage garden – no mention of how it's a vulnerable species, no mention that at peak occupancy the Gardens has between five and eight

per cent of the total population of Grey-headed flying-foxes – so
we were not happy with the unopposed media they were getting.
We started to try and work on how we could use that legislation
to benefit, how we could make submissions – both myself and
Storm are fairly bloody-minded, which we found has been quite
useful . . . We stick at it . . . We use Freedom of Information to
get information that they don't want to release to us. Everything
we do is completely legal. We will foster contacts inside organisations. We will talk to people who perhaps the government
wouldn't expect us to talk to, and through doing that we build up
a much fuller picture of what's going on . . . We do whatever we
can to unsettle that balance so they're not constantly on the front
foot and we're on the back foot. . . .

One of the conditions on the Trust was that they needed to fit
radio tracking devices to sixty female Grey-headed flying-foxes.
Because that's part of a scientific programme, they also need to
get an ethics committee approval for it. The ethics committee
approval that they got states that they can only fit radio tracking
or satellite collars to females that are more than 650 grams.

This was to be a scientifically monitored action. The scientists
needed to be able to track a specified number of flying-foxes so
as to follow where they went after giving up on the Garden. The
scientists were up before dawn waiting for the flying-foxes to arrive;
they had put up their 'mist nets' to trap the creatures. They had to
catch, weigh, check the sex, and fit the collar, doing it all as quickly
as possible to minimise stress. The plan never came off well. It
turned out to be a tough year for flying-foxes and the scientists in
charge of the collaring programme could not get enough animals
of sufficient weight. They said that the programme should be post-
poned. The Garden asked the animal ethics committee to alter the
protocol to allow the programme to proceed, and to the dismay of
scientists and others, the protocols were altered. What was ethically
and scientifically correct at one moment was suddenly no longer
necessary. Meanwhile, the flying-foxes, unaware of the horrors to

come, were nevertheless behaving in ways that amounted to resistance. Tim Pearson explained:

> The real problem is that we have no idea what the result will be. The scientific monitoring programme that was supposed to be an essential part of the dispersal has been handicapped by the fact that many of the bats trapped and radio collared appear to have quickly left the Sydney area, suggesting they were probably just passing through. This means that effectively the radio collaring will be largely pointless as a method to track where the dispersed bats actually go . . . As an aside to this, one of the people monitoring the gardens has commented that many of the bats flying in to the RBG camp pre-dawn are following a very odd pattern – usually they fly in low and fast; many of them she observed were flying very high and then dropping vertically into the camp – we suspect this is the resident bats' reaction to the constant presence of the mist nets after 18 months of trapping. This suggests, perhaps, why no resident bats are being trapped.[31]

So, in spite of all the time, effort and money spent to document the after-effects, the data just weren't there. The expulsion started on 4 June 2012. In a move that bears alarming similarity to the famous decree that 'death solves all problems' (attributed to Stalin), the problem of protecting trees was presented as amenable only to one solution: the elimination of flying-foxes. *The Sydney Morning Herald* put it in terms of combat: 'Botanic Gardens bats given their marching orders'.[32]

Excursion into the Garden

In June 2010 my colleague Natasha Fijn joined me to make a few videos about flying-foxes in Australia. In one segment, Tim Pearson took us for a walk through the Garden and discussed the lives of flying-foxes, plans for the 'dispersal' (expulsion), and measures in place to monitor the process. The video shows a lively and convivial

place where flying-foxes, trees and flowers, and visitors are all wel-
come. Retrospectively, it looks like a glimpse into Eden.[33]

Two years later, on 4 June 2012, shortly before the rise of a stun-
ning full moon, the harassment began again. Later there was a lunar
eclipse, and it seemed burdened with significance, as if the glory of
the moon was being overtaken by darkness as a cosmic comment
on this cruelty. Australia has known many dispersals. In the days of
conquest, the term was a euphemism for killing Aboriginal people.
These days it still seems to convey an intention to disguise something awful. And there can be no doubt that what the flying-foxes
are being subjected to is terror. That is the point. They wouldn't
leave otherwise.

I joined Tim and a few advocate/carers briefly as they documented the sound and chaos. This is not a quiet city so it takes a
lot to disturb Sydney flying-foxes, but as we watched them flying in
desperate confusion, not knowing what was happening, not knowing what to do, there could be no doubt that terror was abroad in
Sydney skies. I was reminded that noise harassment is another of
those words, like cull, that hide an ugly truth. Actually, noise harassment is otherwise known as sonic torture. Like all torture, it is
forbidden under international law when it is applied to humans. The
effects of sonic torture are just what we observed on that cold, damp
night at the edge of the Botanic Garden. Below us we heard the
noise broadcast around the areas where flying-foxes were camping.
In the darkening sky above us terrified creatures flapped clumsily,
their coordination askew, their senses clearly shattered. I couldn't
bring myself to go to witness the early morning horrors when they
tried to return. Tim told me it was even more awful. In technical
terms, sonic torture breaks down the integration of the subject's
senses, creating a state of confused chaos that disconnects the subject from their physical world. No longer able to make sense of the
relationship between self and world, the subject of sonic torture is
a creature lost not only to itself but to its surrounding life world.[34]

Breaking into the life world

Italian scholar Adriana Cavarero helped me think again about the problem of using the language and imagery of warfare to describe contemporary violence. She shows that there are huge problems because so many forms of violence are directed primarily against the helpless. Her examples all concern violence perpetrated by humans against humans, but the general direction of her analysis works equally well with violence against animals. The key point is this: 'violence against the helpless is becoming global in ever more ferocious forms, [and] language . . . tends to mask it'.[35] The masking language draws on images of warfare, but much contemporary violence does not live up to the model of the warrior. Violence against the helpless, violence for the sake of making life utterly miserable and uncertain for those against whom it is directed – this is not warfare. This is something that should be named as a hideous phenomenon in its own right. Cavarero calls it horrorism; she describes actions that 'dismember and disfigure the body, the social relations, the uniqueness of that way of life'.

Is horror new? Not at all, Cavarero says, and yet something is changing. In part it is the scale of violence, in part it is the organised and sanctioned targeting of those who are helpless, and in part it is the wanton revelling in ruining others, their bodily dignity, their life and future. She discusses the totalitarian principle that 'everything is permitted' in the use of force against the defenceless. Here in Australia we have legislation that prohibits cruelty to animals and the purpose has been very clear. *Not* everything is permitted in the use of violence against animals. But when Queensland reinstated the legality of shooting flying-foxes and decided to reduce restrictions on dispersals, it opened the way for an apparently bottomless pit of cruel and vicious action. The announcement was made on National Threatened Species Day, 7 September 2012. It had a political objective as well, targeting 'greenies' as well as flying-foxes. It was a poke in the eye to people who dedicate their lives to defending vulnerable creatures. According to the new law, all four species could be shot; there were quotas for each. Action that had been disallowed because it was inhumane now became permissible. The message was clear: yes, there had to be a permit to shoot or otherwise 'disperse' flying-foxes, and, yes, actions were meant to comply with regulations, but in the absence of any outside scrutiny, and with the tacit approval of local councils for whom 'everything is permitted', cruelty became a matter of local choice.

Cavarero tells us that the language of warfare actually puts a layer of conventionality over actions that are essentially crimes; so let us not forget: actions that would legally have been crimes if the legislation had not been changed are still the same actions. Nothing has changed except that people are now carrying out violence that previously the courts, the legislature, and all humane people had understood to be criminal. In the language of horrorism, people are savaging the bodies, lives and future of those who have no means of defending themselves against this uncontrolled, relentless wounding.

* * *

Until recently it had not occurred to me to wonder about the effects of water cannons and helicopters on small vulnerable creatures. The Queensland town of Charters Towers planned just such a dispersal, using these and other weapons. A petition organised by flying-fox advocates in advance of the action stated:

> All flying-fox camps are full of mothers and babies at this time of the year and whilst many babies are still being carried by their mothers, the majority are too big for mothers to fly with and will be left in the crèche trees at the mercy of the water cannons. Water cannons break bones and helicopters create down drafts that smash bodies and wings.

Charters Towers is a fine-looking inland town in Far North Queensland. It was built during a gold rush and has substantial heritage-listed buildings and a beautiful park in the civic area.

Altogether it is a lovely place to visit, at least on the face of it. Every year Little Reds and Blacks come to visit, sometimes in large numbers. The park in which they camp can become unpleasant for humans when the flying-foxes are visiting in large numbers. In December 2013 residents who had been complaining about the presence of flying-foxes in the park were successful in gaining permission to undertake a dispersal. Advocates and carers came to document the event, and to assist wounded flying-foxes. They were not welcomed. Noel Castley-Wright, a professional filmmaker, produced an excellent documentary about the dispersal and the political hype that surrounded it. The State Premier Campbell Newman is shown calling for smoke bombs, noise and helicopters in order to achieve a dispersal that will put human interests first. In spite of expert advice on managing mutuality, the issue was presented as 'us vs. them': 'We will always stand up for people ahead of bats.' Mr Newman expressed regret: 'No one wants to see these wonderful native animals harmed, but. . . .'

The assault on the flying-foxes took place with local government approval. In no way was it a dirty little secret.[36] This was said to be legitimate: authorised human beings attacked defenceless creatures with smoke, water cannons and firecrackers; they used helicopters to fly low so as to terrify flying-foxes and create downdrafts that break their wing bones. They shot paintball guns; and flying-foxes were struggling to fly with holes in their wings. When individuals flapped around in terror or fell to the ground injured and in shock, the violence-minded human crowd was there to witness. When they saw the wounding and the struggles not all of them shared the same commitment to violence, but they cheered.

Not all the residents took the same view, and one man, John Brophy, offered facts based on decades of living in Charters Towers. The Reds only came for short periods, he said, but they came in numbers. They used to camp in an area beyond the edge of town until some people cut down all the trees. Brophy seemed to regard annual visitations of flying-foxes as part of the seasonal life of the town. He was critical of the dispersal, and he certainly was not cheering.

Is the Charters Towers event over? Not for flying-foxes. Not for those that came back a few months later and were assaulted again. And probably not for the survivors that perhaps have gone to other towns in Queensland. It is terrible stuff to have to stay with for too long, but those who suffer, whether human or animal, don't have a choice. At the very least, we who have not yet been drawn into the vortex of violence are called to recognise it, name it and resist it; we are called to bear witness and to offer what we can.

Like other forms of torture, animal suffering caused by humans is designed to accomplish on an embodied scale what it asserts on an abstract scale; that is – human sovereignty over the nonhuman world. Its intent is to break into the bodies and life worlds of animals. Dispersals go to the heart of flying-fox life: their mental maps, their nurturance of their young, their subsistence, their sex lives, their desire for home places.[37] There are legal constraints on dispersal methods and timing, but here too cruelty is being re-authorised. The Queensland government is changing the rules so that local councils won't have to get permits to disperse colonies.[38]

The torment used in dispersals relies on standard techniques of torture: sonic torture, physical pain in the form of smoke and chemicals, the use of bright lights, and the destruction of home places. Videos taken during a dispersal in the Queensland town of Barcaldine showed a programme of lights, noise and chainsaws. The action was undertaken at night, and as the flying-foxes came home after a night of foraging, tired and ready to sleep, they found that their home had become a place of violence. The cries of juveniles who had been left behind in camp are clearly audible, and the panic of the adults who didn't know what was happening is visibly evident. A few individuals braved the noise and lights to land in a tree. Slowly, inexorably, the people with chainsaws cut the tree down.

Louise Saunders went to Barcaldine to witness and to assist the injured. She described it vividly:

In 2011 I was involved in the Barcaldine dispersal. That was quite an eye-opener. You can see these townships are so important to flying foxes. They are an oasis of green in a sea of cotton or dry, awful land that offers up nothing for the bats. And you have all these dry years where there's no nectar. Traditional roosts are either no longer there, having been bulldozed, or the bats are not allowed to roost there anymore. So where are they going to go? Into the townships. . . .

But to see 60,000 Little Red flying-foxes in one quarter-acre was pretty amazing. I really did feel for the people. Because the lady had built this beautiful forest in her backyard and it had been obliterated. The bats had absolutely annihilated it. All the palms were shredded, there were just trunks and dead leaves hanging down. All the trees were just bare branches. It was quite awful. I could see how they felt. . . .

The very first night the local people went in at five o'clock and just started annihilating what was left of the backyard vegetation you can see in that footage great big gum trees coming down with bats hanging in them. The next day they were going to come in and annihilate the side and the front vegetation. I said 'don't do it, you've done the backyard, you don't have to touch the rest, wait and see what happens'. Thankfully they left that vegetation. The next morning they were doing all of the dispersal noise again, the fire sirens . . . Incredible noise, just awful noise, and the poor little things, you could see them coming in after flying all night doing circles and circles, working out what they were going to do. . . .

There were about 60–100 that stayed behind and they were mothers calling babies. The dispersers didn't let the mothers stay to catch up with their babies. They were just dispersing every-thing. We had a few babies left behind. They were struggling. They just couldn't fly anymore. They just knew that this was home and they had no idea where to go. You see a grown woman with a ten-year-old daughter and she's using a leaf blower, just blowing this little baby. I just said to her, 'That's a little baby flying-fox and it can't fly anymore.' And she just kept on doing it. And I said, 'You've got to stop now!'[39]

By morning, all that was left was a desolation of dead trees and disappeared flying-foxes.

In the long run, dispersals shift 'problems' from one place to another, and often the result is worse than the original, both for humans and for flying-foxes. The distinguished scientist Les Hall asked the crucial question in regard to dispersals: 'what have we learnt in the last twenty-five years?' His answer was that long-term observations suggest that moving a camp just moves problems to other areas, 'and the whole process starts all over again with a new lot of players'.[40]

What does hatred want?

Hatred is integral to persecution. Among other things, it helps sustain boundaries (such as 'us' and 'them') in a mode of 'either-or' rather than in the symbiotic, generative mode of 'both-and'. The logic of either-or is built on danger and fear; linked to torture and expulsion, it means 'they' have to go.

In an important study, *The Ideology of Hatred*, Niza Yanay writes that while it is almost pointless to ask what hatred is, another question is well worth analysis: 'what does hatred want?'[41] If hatred is to be effective, it must be stimulated, sustained and reinforced. Fear is its ally. Vilification is a prime tool for generating a shared social experience of both fear and hatred.

On the political side of many of the stories of expulsions, the decision was at least as much about party politics as it was about the animals. It was anti-Green, pro-rural and against anything that might alienate rural votes. So, hatred wants votes, and votes are a means to an end . . .

Hatred wants political power

And then there is the scapegoating. Flying-foxes are among several species bearing the brunt of anger and fear amongst primary producers. In the face of climate change, droughts, uncertain global financial markets, international regulations, land degradation, and long histories of ill-informed practices, animals are blamed, and killed . . .

Hatred wants security, and failing that, someone to blame

The 7 September decision (announcing the resumption of shooting in Queensland on National Threatened Species Day) showed a set of sovereignty issues working across several scales: state vs. commonwealth; rural vs. urban; and the right of private property vs. any legislation emanating from anywhere. Questions of sovereignty are integral to all this death. Foucault's analysis of biopower probes the exercise of sovereignty as the right to decide who lives and who dies within a system ostensibly devoted to the improvement of life. Racism and colonialism enable sovereignty to manifest itself through the 'destruction of bodies and populations'[42] . . .

Hatred wants sovereignty

Politics (votes and power), sovereignty and security converge in a wildly toxic mix of danger and retaliation, often expressed vigorously and publicly by Bob Katter (MP). His fulminations on television news about the need for more killing were described as a declaration of war. Standing in the garden of a property that was quarantined because of Hendra virus, he shouted: 'They have to die.' Moving on to the issue of people's right to shoot flying-foxes in their backyards, he said: 'They're the greatest possible danger to human life, to your wives and children, and the government says you can't remove that danger. But like damned hell!' This comment was elaborated in another report in which he criticised both the federal government and the state government: 'Now these two warped groups of people honestly believe that these flying-foxes have the right to your backyard, but you don't have the right to your backyard.'

This inversion of victimhood has the curious effect of implying righteousness.[43] Human perpetrators of terror understand them-

selves in a wildly contrary way: they are the victims of terror, and
all they want is to launch a counter-attack in a field of violence
initiated by the enemy[44] . . .

Hatred wants to be the victim, and to be righteous

Hatred wants enemies, brings them into visibility, and inscribes its
power upon and into their bodies in modes of terror. 'Systemised
terror creates a relentless climate of anguish.'[45] These vivid words
evoke the lives and deaths of flying-foxes as they are electrocuted,
shot, expelled by all manner of means, orphaned, starved, impaled
on barbed wire, and otherwise subjected to direct and indirect technologies of terror. And still, there is righteousness . . .

Hatred wants to be both violent and innocent

Hatred thrives in this potent mix of fear and desire. I asked the
experienced advocate about how she understands the hatred that
she has battled against. Her response was thoughtful and, I thought,
rather generous. She has had a lot of contact with these people, and
understands a range of fears, annoyances and perspectives:

> I've grappled with trying to understand it over the years.
> Sometimes I think I understand it and other times I feel com-
> pletely nonplussed about it. Flying foxes really do symbolise
> something far beyond what they represent in reality. I think it's
> because they trigger a number of things, that perfect storm sce-
> nario. It's the property rights thing. They've become something
> farmers have to fight. So they represent greenies' growing power
> to influence what people do on their own land. And it's because
> they're in town, in urban areas as well. And they're annoying . . .
> So there has always been a lot of anti-flying-fox rhetoric – it's
> been there ever since colonisation – but it is just so exacerbated
> by the rabies and disease thing. As we know, people have such
> an unreasonable response to risk . . . They don't necessarily think

about it in terms of what risks are, they think just that they can be killed by this disease.[46]

Depersonalisation

Janina Bauman, a survivor of the Warsaw ghetto, wrote that the 'cruellest thing about cruelty is that it dehumanizes its victims before it destroys them'.[47] Is there an animal counterpart? Are flying-foxes depersonalised? There is a triple answer here: the process of depersonalisation requires that flying-foxes be humanised since one cannot dehumanise that which was never human; then they are dehumanised, and then de-animalised or depersonalised, finally being deprived of their life world and their way of living meaningfully in the world.

First, these animals are humanised as enemies. Their presence in people's backyards is spoken of as warfare. They poison the water and spread lethal diseases. They kill children. They cross boundaries of private property, and attack people's safety, security and prosperity. The rhetoric of hatred humanises flying-foxes by constructing the identity of an invading enemy.

The second part of depersonalisation is vilification, with the category pest being strategically deployed. As we have seen, a large research and technology industry exists to kill or otherwise control pests.[48] The third step in depersonalisation is de-animalisation. Flying-foxes, you will recall, are participants in biocultural loops that circulate across generations and geographies, and involve trees and shrubs, home sites, long-distance travel, their own social lives and much that we humans will never grasp at all.[49] The category 'pest' discursively eradicates that life and authorises physical eradication as well. And so de-animalisation goes on in cascades of violence, tree by tree and camp by camp, breaking into the meaning of life in relation to place, breaking into the bodies of lost juveniles and desperate mothers, cutting into the future of flying-foxes, and forests.

* * *

Emmanuel Levinas's essay 'Useless Suffering' shows that suffering eradicates meaning within the one who suffers – it takes over the whole being, destroying not just the body but the being.[50] Evil, he goes on to say, is the deliberate infliction of suffering and thus the eradication of meaning in the others. This strong sense of evil in relation to animals depends on the fact that there is a meaning world, or life world, to be attacked. One of the many big questions is this: how is it that we live in the midst of evil even though, for so many of us, our deep desires are to live symbiotically in multispecies alliances and collaborations?

Many of the decisions to disperse and/or kill were authorised in a language of managerial oversight in which questions of cruelty were cunningly deflected. Recall the words of the Minister for Environment, Andrew Powell: he asserted that the killing would be as humane as possible, and that he himself sleeps well at night. Likewise, Queensland Premier Campbell Newman, calling for smoke bombs and other forms of violence, assured television viewers that 'no one wants to see these ... animals harmed ...'. We see here the 'non-performativity of anti-cruelty'. I am paraphrasing Sara Ahmed's inspired analysis. Non-performative acts offer the impression of accomplishing something, but the statement of action takes the place of real action, and may even get in the way of real action.[51] Premier Newman's insistence on violent dispersals, coupled with his bland assurance that no one *wants* this, fulfils the core function of non-performatives, that is, of '*not* bringing about the effects that they name'.[52] In this case, the non-performative both authorises cruelty and denies responsibility.

In politics and in research careers, the language is usually extremely muted. Zygmunt Bauman has studied these matters deeply, and he tells us that one of the main ways the modern world narrates mass death is as '*creative* destruction, conceived as *healing surgical operations* and perpetrated in the course of paving the road to a ... harmonious society ...'.[53] The dynamic we've seen with flying-foxes is that it takes both: the managerial surgery, and the wild and violent vilification. There will be bloodshed, torture, grief and unimaginable suffering. There will be calls for warfare and eradication. And at the same time, somewhere nearby (probably not far from a TV camera), a person in a position of power will smile and tell us that . . . it's all okay.

Emptying Earth's life world

The wider biocultural import of all this violence, however it is talked about in public, is that it slashes into the shimmer of ancestral power, breaking up the intergenerational waves of nurturance, diversity, fecundity and joy. The gleaming enthusiasm for life, so characteristic of flying-foxes, is being replaced with terrible negations. Considering the Sydney Royal Botanic Garden expulsion, the ideas it embodied were almost entirely negative. They augur emptiness and ongoing trauma. Life's great desire is Yes! What we saw in Sydney, and what we have seen in so many places, is a raging negativity:

- Zero tolerance: there can be no accommodation with flying-foxes.
- Zero compassion: never mind that the females are pregnant, and are being subjected to enormous ongoing stress, and that this is a maternity camp to which they intend to return.
- Zero planning: no one knows, and no one can know, where the flying-foxes will go next, but most options are likely to be worse than where they are now.
- Zero time-limits: the Sydney dispersal for example was designed to extend across thirty years, generating large-scale intergenerational impacts.
- Zero bottom line: the cost of expelling flying-foxes seems to be greater than the cost of putting in infrastructure or other management programmes that would enable peaceful co-existence.
- Zero precaution: spillovers of terrible diseases are instigated by stress; taking people's lives seriously would entail taking flying-foxes' lives seriously.
- A total absence of awe: it is hard for me to imagine, but apparently

some people are immune to the beauty and mystery of these creatures. Looking up into a blizzard of flying-foxes, many, many people feel close to paradise. In contrast, the hatred or fear which prevent some people from experiencing all this beauty and enthusiasm must be strong, and indeed announces a wilful turning away from participatory life.

This is our time of unmaking. And as we unmake the life world that brought us forth, the losses keep on expanding. Everyone's 'paradise' is being shattered.

9   Fidelity

'Yes!' is the great powerhouse of life on Earth. Life moves, it acts, it comes bursting forth. We know this because we ourselves are inside the great Yes![1] Yes! asserts the value of life over the powers of destruction, and flows through living creatures in all their fidelity to their own way of life, their new generations, their mutualisms, and their flamboyance. Yes! expresses passion and participation. It opens our attention towards an ontological-ecological terrain of mutual life-giving: not 'self-preservation' in the singular, but the care of relationships and connectivities; not self-enclosure, but interdependent waves.[2] For humans, saying yes to life is a profound ethical choice. It is an embrace of the living world, a grateful response to the gifts of life, a pledge of solidarity with Earth's way of becoming, and a commitment to witness to the work of life. Responding with fidelity, we are inspired to remain radically open to others and to the future and to the past.

No one is alone

Hobbles Danaiyarri's account of creation started with the ground and with action: 'Everything comes up out of the ground – language, people, emu, kangaroo, grass. That's Law.' Within this Earth-based

Law, everything is either contained within or is emerging; action is ongoing, and it is open-ended. The containment aspect of Law is fundamentally significant because the world (organic and inorganic) is open to becoming and co-becoming. Not everything that could ever happen has already happened, and because we do not know what Earth holds and has yet to bring forth, we are both in it and potentially drawn to it again and again. Indeed, those who are attentive are frequently astounded by novelty – not only the novel abundance of 'things' but the proliferation of connectivities. In our daily lives as well as in heightened events, we witness the coming forth of the continuities and the unpredictabilities, both of which mark so much of life.

Hobbles's account of everything coming forth does not offer a Garden of Eden. 'Everything' is not human-centred. Dreaming Law brings forth all that is, and some of what is brought forth in creation is tough for humans: poisonous snakes, for example, or stinging ants. If earth life were designed for human comfort, these creatures might not exist. Fidelity to creation Law is not about comfort. In Hobbles's world there is medical knowledge; humans and others are not without remedies and living well does not entail deprivation, but nor does it offer invitation to judgement. Rather, creation calls for living within modes of deep, connective fidelity. None of us knows what will come forth – for ourselves, our family, our tribe, our species – but within Earth, the promise of life endures.

* * *

A great many of us western-influenced people find it difficult to maintain an acceptance of Earth life. Questions abound. In previous centuries a framework already existed; the meaning of life and death had divine origins, and much could be explained by reference to the sacred or, alternatively, to the active existence of evil. Indeed, many of the world's traditions dedicate themselves towards understanding suffering, and they develop deep connections between the sacred and the hardships of life. In the secular West, however, matters became more difficult. A pivotal figure was René Descartes

(1596–1650). There is an element of tragedy in this story. Descartes aimed to move away from the religious violence of his time, and to approach religion with reason rather than faith. He held a rigid adherence to a mode of thought he characterised as reason, and he denied sentience to all other creatures. All knowledge, he was saying, was the product of the human mind exercising reason, and thus humans stood over and apart from 'mere matter'.

Many thinkers have taken up critiques of Descartes. Most relevant to the ontological-ecological terrain explored here, contemporary eco-philosophers have shown many of the problems of holding the nonhuman world to be composed of 'mere matter'. The violent separation of human reason from all else produced a terrible 'de-realisation' that leaves no place for mutualisms. That famous insight 'I think therefore I am' opens a weirdly shadowed vista: if only humans think, then only humans could be understood fully to exist. We are left lonely and bereft. In Freya Mathews's words: 'if the world cannot be shown really to exist [since other beings do not "think"], then it can scarcely be shown to matter in its own right'.[3] In contrast, philosophical traditions in many parts of the world, and contemporary eco-philosophy at this time, encounter in the nonhuman world a region of active, indeed captivating, complexity and consciousness. In Val Plumwood's words, what grabs us is 'the incredible, infinite complexity of the real earth story written in the rocks and in the bodies of living things, species diversity and evolution . . .'. Rather than a world of inert matter, we live within creative, active matter. The challenge is to recognise that fact.[4]

Lev Shestov was one of the few early twentieth-century philosophers to develop a critique of Descartes founded not in more elaborate reason but in love of life. His passionate commitment to birth, life and the joys of the ephemeral also included a dark perception of the consequences of reason without love. Shestov abhorred the idea that G-d could be made answerable to human judgement. His love of the mysteries of being and becoming led him also to abhor the idea that human reason could be the sole measure of all that is and all that ever can be.[5] Shortly before World War II he wrote eloquently against the West's commitment to dualism and the denigration of Earth life. In his prescient words: this commitment 'would poison the joy of existence and lead men, through terrible and loathsome trails, to the threshold of nothingness'.[6]

* * *

An inevitable question arises about why suffering exists. Many traditions address this question, and the responses vary enormously. The subject is so huge that I must hold my focus specifically on multispecies life, and on the fact that suffering articulates a call into ethics. Where suffering and ethics meet, at just this conjunction we are called to witness and to respond. These issues are big, but the core point brings us back to the fact that while suffering is an inevitable part of life, the sources of suffering are not neutral. The suffering that is usefully and inevitably part of life is not the same, existentially, as the suffering that is wilfully imposed on others in order to make their lives miserable or even impossible.

Levinas encapsulated his insights into evil by pointing to 'the justification of the neighbor's pain' which, he said, 'is certainly the source of all immorality'.[7] This insight is not a guide to specific action. Levinas is not seeking to prevent the unpreventable. Rather, he tells us that evil consists in making others expendable, promoting or justifying their suffering by reference to what is claimed to be a greater good.

No one is ready

'Gaia' is one of the twentieth century's terms for the biosphere, brought into contemporary thought by James Lovelock. His analysis focused on the whole Earth system, and he proposed that the functioning of this dynamic planet is not a matter of pure chance. Rather, he held that 'early in life's history living and nonliving matter became entangled as *a single entity* within which organisms themselves may have been shaping conditions to their adaptive advantage'.[8] The bombshell impact of this analysis was to totally reject theories of mere matter, bringing connectivity, agency and adaptability to the fore in seeking to understand the biosphere as a living system. According to two of the key thinkers in this field: 'After 400 years of being virtually shelved by dominant mechanistic and reductionist perspectives', the concept of Earth as a living entity has become integral to interdisciplinary Gaian research.[9] Key concepts are openness and connectivity: Gaia analysis shows a 'world of open, flowing, genetically and thermodynamically connected forms'.[10] Connectivities are essential, and they are everywhere, but they are not chaotic. Gregory Bateson famously asked: 'What is the pattern that connects?'[11] Patterns are organised variations: they require both difference and sameness. The mutualisms and co-becomings that sustain life depend on the fact that they are not random, and so the connectivities become productive by moving the flows of life across patterned boundaries.

In popular culture Gaia has often been represented not only as female, but as a mother with kind and generous intentions towards the life she brought forth. Such thinking was always wishful. The biosphere desires life, we know that, but there is no indication that this favours any species over others, and it has no preference for humans, as far as is known. The 'mother' story, like Descartes's pure reason, and like traditions of divine favour, is a convenient gloss exempting us from fidelity to responsible participation in the inherent mutualisms of our multispecies world. These pleasant stories stop conversation just at the point at which serious thinking would be required, pushing difficult issues across an arbitrary boundary into a realm that, under dualistic thinking, cannot respond. The status quo thus purports to be inevitable when in fact we live in the midst of ongoing social and cultural constructions.[12]

Perhaps because of these pop-culture connotations, Gaia theory is often called Earth Systems Science. Lovelock's emphasis on self-regulation appeared to offer an image of relative stasis by emphasising Earth's self-regulating capacities. Interestingly, the last 10,000 years or so have been a period of unusual stability in Earth's climate, particularly in the temperate zones where horticulture and settled 'civilisations' were forged. Stability became part of ever-expanding human cultures that depended on it for their agricultural success. Calendars offered visual representations of regular, stately progressions of seasons. Such stability, although never perfect, was foundational to many of the philosophical ideas of the eternal and immutable.[13] I do not mean to suggest that environment determines culture, but there are patterns of consistency between human thought and ecology that are particularly helpful for us to investigate now that everything is changing. Aboriginal Australians, for example, living in the driest and most unpredictable continent, well understood the dynamics of change and unpredictability. At the same time, their understandings of Earth life included a form of intra-active regulation. The wrestling antagonism between the Sun and Rain is an example. Each force pushed its own climate regime – dry and wet – and each threatened to dominate. And yet, each was checked by the other and brought back into balance. Stability was not a stately procession of seasons, but rather an intra-active, ongoing tussle in which communication was essential. Flying-foxes were keystone communicators as well as keystone pollinators, and their calls for the Rainbow to respond were calls for balance (see Chapter 4). From an outside perspective these relationships can look stable and calendrical, but from the inside this self-regulation is actually about intra-action and ongoing, ever precarious coming forth.

Evidence of contemporary climate change is radically disrupting earlier perceptions of stability, and now there seems to be new mainstream recognition that major changes are already in force and are increasingly powering themselves into ever more dramatic changes. The concept of feedback within systems is essential. Self-regulation depends on negative feedback; this means that a system, such as the Earth system, corrects itself as it orients itself towards its goals (life, in this case). But when change becomes extreme and rapid, it may outstrip the capacity of the system to correct quickly and effectively.[14] Sun, rain and flying-foxes are enmeshed in relations of negative feedback. In contrast, positive feedback loops, or runaway feedback, are drivers of accelerating changes leading to more change. The geological record tells us that negative feedback loops are re-established in due course, and that life goes on in new and dramatically diverse forms, but while we are in the midst of unknowable, unimaginable, runaway processes, a four-billion-year history of life's capacity to thrive may be of small comfort.

Our current time of change is complex: extinctions, habitat loss, toxins, plastic pollution – a comprehensive list of damage leading towards death of existing life forms would be very long. Human efforts towards addressing the factors that promote runaway systems remain on a far horizon. As climate science and a great deal of ecological science have radically altered our perceptions of stability, it becomes clear that rather than self-regulating towards equilibrium, Gaia, or the biosphere, is now doing things whose power completely escapes our understanding, manifesting 'positive feedback loops that terrify those who study her'.[15] Gaia does what Gaia does. Human desires to control, placate or harness Gaia energy hold open some possibilities for outcomes other than disaster, but when it comes to the ecosystems and to the biosphere, we are not in control. In Isabelle Stengers's forceful words: 'No one is ready for what's coming. It is beyond all of us.'[16]

* * *

How is it that we humans, so many of us, seem to be helpless, or even resistant, to addressing in tough and coherent ways great challenges of our time? Why, when we know so much, are we so unready? Isabelle Stengers asserts that we have been under a spell woven by the great powers of capitalism, corporatism and their wide network of support systems. The power of these forces goes beyond the usual paradigms of social problems and social solution. She writes of the seemingly overwhelming power of current dominant structures in science, philosophy and society that 'crush our hopes' of ever getting outside them. It is as if they had the one true account of reality against which it is useless to struggle. Stengers asserts that these powers capture us, in fact subject us, to sorcery. Ordinary resistance will not extricate us. True, we need to dedicate ourselves to working against the sorcery or 'spiritual capture' that has taken hold of us; we have to mobilise the usual methods of politics, persuasion and advocacy, and these are not trivial. But, at the same time we need to develop antidotes that, working against sorcery, would contribute to a new politics. We would then be drawn into practices that enable us to live *for* the world, and to believe *in* the world.[17] The incapacity to act calls not just for an end to standard politics of winning and losing, but for actual experience that takes us outside the ensorcelled domain of rational means and ends. One antidote Stengers offers is, actually, this living world – the world we are rapidly wrecking. This is to say that we are called to direct our thoughtful capabilities towards the living world where other mindful, active and intra-active beings are living in ways that are not ensorcelled, even though powerful spiritual capture bites into our capacity to recognise and experience multispecies ancestral waves of power.[18]

Waves of desire

Deep within the biocultural world of a flying-fox camp, individuals pursue lives of impassioned, but generally impermanent, connections. I was surprised to discover that some clichés work across species. A female flying-fox's work is pretty well never done. This does not mean that it is drudgery, however. Sex is far more extensive and mutually active than one might imagine given that the biological outcome is a single pregnancy. After a few months of sex play and conception there are about six months of pregnancy. During this time the females have their freedom to travel, explore and chase blossoms. They experience and enhance their own individuality along with engaging in unexpected social encounters with others. When it is time to give birth they return to the maternity camp, and the next few months are dedicated to nurturing, teaching and, ultimately, weaning their little one. All this work of nurturance requires passion, dedication, worldly knowledge and intergenerational knowledge. After the youngsters have become self-supporting, the mother-child bond that was so intense seems to disappear. Females no longer have those deep responsibilities, and they turn their attention to the males who have, for some time, been gearing up, culturally and biologically, to mate. The following months of sex play involve enthusiastic pleasure.

Each intersubjective relationship appears to be both intense and short-lived. Babies go on to live their own biocultural lives. Sexual partners seem to be transient, and there is a great deal of choice. The constants in a diverse field of desires include the ephemeral blossom feasts which, although they come and go, have so far always come again. Loss of blossoms is having the effect, as discussed, of pressing flying-foxes to turn towards other foods, but the evidence shows a continuing passionate desire for the preferred foods. Even more deeply constant are the home places. Until the major assaults of land clearing and dispersals began seriously to break site fidelity bonds, home place was deeply stable. Site and blossom fidelity are the two poles around which the transient lives of individuals depart and return.

To allow ourselves to be drawn into these realms of enthusiasm and desire we have to consider pleasure, nurturance, love, play, communication. These qualities of touch and connection include the caress, the hug, the voice; the mutual pleasures of contiguity, intimacy and eros. We learn of them succinctly through thinking about the long, slender flying-fox tongues.[19] They are loaded with taste buds and covered with small papillae that give a softly textured feel. They are central to every aspect of life, from birth to feeding, to grooming of self and others (a major daytime activity), to sex and to play, as well as to the nightly work of eating.[20]

Flying-foxes' annual cycle includes several months during which the adults organise and conduct their sex lives; during this time, the females become fertile, and the males reach their annual peak of sperm production. Although sexual activity heats up during these months, some sexual activity also takes place well outside of the months of intensity.[21] Flying-foxes mate during the day, and occasionally they engage in 'sudden bouts of mass mating'.[22]

The male-female relationships are defined as polygamous, a term developed for describing human marriages, but in this wider usage simply means that every male seeks to attract several females. The most successful males, in competition with other males, are those with more than one partner. Evidence suggests that they cluster in the centre of the camp, leaving less successful males at the outer edges. Males become sexually active after about two years of life, and generally those with only one partner are younger. There are, of course, also younger males and females who stay in the area for the food and sociality on offer but who are not yet sexually active.[23]

There is a time during February, March and April when the males stake out their mating territories, marking them by rubbing them with special scent glands located on their shoulders, and defending them from other males. They groom themselves and display their genitals, enhancing the display by licking them. A flying-fox penis is an impressive sight when on full display, as one would imagine give the capacity for self-licking. The aim is to attract females. Once a female (or perhaps several) joins the male in his camp, the male's attention turns towards enticing the female by licking her genitals. Not only do flying-foxes miss out on sleep during the mating season, the males also curtail their foraging. During this period, the males and females fly out at about the same time every night, but the males return after a couple of hours in order to continue to defend their mating territories, refresh their scent marking and screech out their 'territorial cries'. In contrast, females continue to feed through the night.[24] These mating sites are places of deep fidelity for males. Numerous studies show that they return to and defend successful mating sites with great obstinacy. Early attempts to disperse flying-foxes from the Sydney Botanic Gardens inadvertently produced evidence of this fact: those who would not leave, or were first to return, were males with established sex sites.[25]

Copulation involves pre- and post-licking.[26] Researchers in China made an explicit study of genital licking, presented in the attention-grabbing paper 'Fellatio by Fruit Bats Prolongs Copulation Time'.

Min Tan and colleagues state that theirs is the first large-scale observational study of oral sex in nonhumans, and they argue both for evolutionary advantage and for pleasure. The research was conducted with the short-nosed fruit bat (*Cynopterus sphinx*); there are no comparable studies of Australian Megachiroptera. The China study showed that 'females . . . are not passive during copulation but rather communicate with the male, in this case by licking his penis'. The authors found that females were licking their partner's penis during intercourse as well as prior to it. 'The bat penis contains erectile tissue . . . which is similar to that found in primates and humans. If the erectile tissue is stimulated during copulation . . . it will increase the rigidity of the penis, and maintain the erection for longer.'[27] These scholars proposed that pleasure was almost certainly a part of a female's decision to use her tongue. They further suggest that it may increase the chances of conception and is likely to reduce chances of STDs.[28] It is clear that the tongue is an extremely significant part of sex amongst many species of flying-foxes,[29] and while evidence from Australia is not so detailed, it does indicate generous levels of licking attention from males to females and vice versa. It is perfectly reasonable to suppose, as the authors of the China study do, that licking involves pleasure.

Conception takes place late in the period of sex play, and there comes a time when this phase of the annual cycle finishes. The time of departure coincides with winter blossoms, and for ecological reasons it is an appropriate time for dispersal. This is the freest time for females, and on the limited evidence we have, it is clear that they travel and explore, expanding their knowledge of geography, other camps and individuals, and sharing in the news that seems to be so central to flying-fox life. Being on the move does not exempt them from human animosity, as we saw with the female who travelled up and down much of the temperate and semi-tropical east coast chasing blossoms, only to end up in a place where mass flowerings of the preferred food had led to large flying-fox assemblies, and then to massive human rejection of the presence of flying-foxes (see Chapter 5).

Closer to the time of birth, pregnant females return to their maternity camp and over the course of a few week thousands, or hundreds of thousands, of babies are brought forth. Flying-foxes give birth while hanging upside down; gravity is not helpful. The mother needs to ensure the baby does not fall, and so she holds it and gets it attached to her quickly and decisively. The birthing mother licks her vaginal area in preparation for birth, and she licks her baby to clean it. She brings the baby to the nipple, and for the next few weeks the baby stays right there, pressed against its mother's belly and holding on to her fur with its feet, keeping her nipple in its mouth, and receiving warmth, protection and nourishment. The small milk teeth are angled to enable the baby to hold on to the nipple while the mother flies out for food at night. After about three weeks the baby becomes too heavy to be carried in this way, and so there comes a time of separation. The baby is left in a crèche area, while the mother flies out for food. When she returns in the early hours, she flies around the crèche seeking her baby's unique smells and calling out in the communication register that only the two of them share. When they reunite the little one reattaches to the nipple and suckles by day. Like most mammalian babies, little flying-foxes don't want to let go of their mother and the nightly separation may become a struggle. Hall and Richards describe the scene: 'Young bats will cling desperately to their mothers' nipples with their curved milk teeth, and hang on to branches with their toe claws when their mothers try to fly away. As a result, the length of a nursing mother's nipple is elongated.'[30] I imagine the dynamics: the baby's demands and fuss, and the mother's firm but possibly exasperated efforts to detach her little one and get on with the work of finding the food that will nourish both of them.

By the time the youngsters are weaned, males are well into the work of setting up their sex camps and seeking females. The cycle begins again with a few months of play and pleasure.

Somewhere along the continuum between grooming and sex, nurturance and maternal care, there is play. According to a study of *Pteropus livingstonii*, play includes 'prolonged gentle wrestling,

holding, mouthing, biting, and genital sniffing and grooming . . .'.[31]
What with all this licking, suckling, playing, feeding and grooming,
we may conclude that flying-fox tongues work hard during the
day. And then night falls and they go out to feed. The significance of tongues for flying-fox interactions amongst each other is
matched by the work those tongues do for other species as well.
Think of the meeting place where the tongue laps up the nectar,
and where the tree opens its blossoms. Just there, at that meeting of tongue and blossom, life happens: there is an interspecies
kiss of great eloquence. 'The kiss brings to attention the entanglement of response and reaction without dissolving those who kiss
into a pool of sameness: kissing is of the edges, of contiguity, not
continuity . . . The kiss . . . finds it difficult to specify' who is the
giver and who is the receiver.[32] Kisses of life continue to move in
the world. At any given moment, there are perhaps two individuals
kissing, or licking, or nectaring, but the story does not stop there.
From creature to creature, flying-fox to tree, and from creature to
atmosphere, tree to air, and from atmosphere to creature, oxygen
to all who breathe: this convivial symbiosis is produced through
heterogeneous ways of being amongst creatures who need each
other, and who seduce and sustain each other. Participation, when
understood in the mode of the kiss, is joyful both in the giving and
the receiving.

Everyone is open to others who care

Traditional politics, the politics that we experience daily, is founded
in the capacity to exclude others, to marginalise and banish them
from domains of ethics. The new politics Stengers advocates concerns 'who is entitled to speak, and on what grounds when *the question of our common destiny is at stake*'.[33] This is ecologically open
in its assumption that we do share a common destiny, and it brings
us back to the fact that no one is ever alone. Sooner or later our
common destiny is always at stake. This is the deep fact that politics as usual seeks to deny. No matter how stringently traditional

politics seeks to close off connectivities and banish those it deems to be outsiders, peace-seeking alliances bring us back to our more fundamentally shared condition as participants in waves of ancestral power.

A large part of the power of everyday politics is the assertion that it is appropriate to sacrifice individuals to the requirements of the 'greater good'. This elusive concept often takes a cost-benefit approach to measure the value of individual lives, and if sacrifices are called for, others will pay the price. Against this vile willingness to sacrifice others, carers, advocates, and many others, exercise an ethic of refusal. In refusing the blandishments of death for others as a quick fix for human problems, carers show us that an ethic of refusal will never allow the integrity and beauty of these 'dramas of encounter and recognition' to become fodder for the justifications of deathwork.[34] Human participants in these action alliances refuse to justify the suffering of others, refuse to abandon others, and refuse to translate the work of care into a politics of expediency.

The faces of others, their calls to not be abandoned, are the bedrock of these alliances, and while accepting that there may be no future for those who are being exterminated, this fact does not erase responsibilities. Tim Pearson explained his commitment despite the fact of the probable extinction of Greys:

> My friend Heather says 'I don't know why you don't give up.' I think the main reason I don't is that while I fear that a lot of what we're doing is futile, I couldn't live with myself if I didn't keep trying.

The carers' strong work cuts across death and, while ever aware of the perils, does not for the most part become focused on probable future outcomes. What matters is the actual lives of actual living beings: their individuality, their suffering, their beauty and their social bonds, their love of place, and their cultural world, which we will never really understand. Carers and flying-foxes belong to episodic alliances that are always forming and re-forming: unpredictable, incalculable, shifting, sometimes unstable, and yet responsive, committed and courageous.

* * *

Alongside these structured intergenerational formations, there is also a region of encounter and engagement, where a human and a nonhuman arrive at a commitment to a unique bond of mutual life-giving. This individualised elective affinity is similar to Aboriginal kinship but arrives as sparks of witness and commitment, rather than being deeply structured across generations. Unplanned and arising in encounter, elective affinities become formative of enduring bonds of life-giving and solidarity. These relationships bring humans and others into bonds of focused, enduring solidarities with specific others. Responsibilities do not go everywhere. What matters is that everyone and everything is already within a world of responsibilities, and no one is left without others who care.

Carers, in my view, can be understood to have been called into a kind of totemic kinship based on their commitment to the bonds of mutual life-giving and enduring modes of solidarity that characterise Indigenous kinship. Elective affinities open the way towards better understanding the depths of love and commitment that keep people working, often under extreme and deeply distressing circumstances. As we saw, vulnerability is part of the story, and joy is also accompanied by tears, grief and the need for time out.

When I met Louise Saunders in 2010 she was stepping away from the huge load of responsibility that she had carried in the organisational side of Bat Conservation & Rescue Queensland. Her words always resonated with love and, as she explained, it was love that brought her to flying-foxes. She had been a botanical artist for over thirty years. Her work is scientifically informed, totally accurate and infused with the love that for Louise arises when one pays close attention to nonhumans. Each work is almost mysterious in the way it combines total accuracy, which itself is a form of respect, with a captivating tenderness that communicates something deeper and

Figure 9.1 Flying-fox generations
Source: © Nick Edards.

somehow more urgent even than respect. We are drawn into awe by the seriousness of the tenderness of her work.

I knew nothing about bats when I came to Queensland. That was in '93. My sister took me on a batty boat cruise up the Brisbane River . . . I have always loved animals and nature. I paint it. You have to love nature to do what I do. My paintings are hopefully a window to nature and inspire people to protect it. When I went on the batty boat cruise and learned about bats, I thought: 'I could do this.' Because I'm working at home painting. And before long I had baby bats hanging off my shoulder and I was still painting away.

And then I got a position at Couran Cove at the resort as the artist in residence. I was there for four and a half years. But I still took care of bats while I was over there. And that was fantastic because all the patrons would come through the art studio and I might have a baby bat in a basket in the corner, and showing people how to feed it, it was changing attitudes. So it was just a really lovely adjunct to being a natural history illustrator.

With time, Louise became ever more involved in the organisational side of rescue and care of flying-foxes. She didn't talk much about hope, she kept returning to themes of compassion and love. Along with being intensely moved by suffering, she was always keen to offer praise. Speaking of flying-foxes' role as pollinators, she wrote: 'Australia wouldn't be Australia without koalas and Eucalypt forests. Both are in danger because their guardian, the flying fox, is in demise.'[35]

In her quiet way, Louise brought a key idea into the ongoing conversations about flying-foxes. I love the way that her term 'guardian' turns potential pathos into mutuality and affirmation of connectivities. Guardianships can be mutual, cross-cutting, rich in connectivities and open to multispecies participation. For Louise, this is what life is about – mutual flourishing, mutual guardians. Louise states it as fact, and it is, but it is also an observed vision of a world in which cruelty has been marginalised and shared care has pervaded all life.

And that the kiss of life includes us, too

Eucalypts and flying-foxes come together every year with beautiful timing and exquisite generosity, giving each other great kisses that bring forth new generations of life. Let us consider the lush, extravagant beauty, flamboyance and dazzling seductiveness with which Eucalypts say 'Yes!' They burst open sequentially, even patchily. And when they burst, every twig says 'Yes! More!' – more buds, more flowers, more colour, more scent, more pollen, more nectar – more, and more, and all that can be conjured from within the tree to reach out into the world with this great, vivid, multisensorial call: Yes!

For their part, flying-foxes respond to the call of Eucalypts and Melaleucas with their own 'Yes!' They sense this great call, and they come racing to the blossoming trees. Their 'Yes!' includes their long tongues that are perfectly adapted to sucking up nectar, and their delicate whiskers that pick up pollen and distribute over 70 per cent of it intact. They carry the future of Eucalypts on their furry little faces; and across the patchy and increasingly fragmented landscapes of contemporary Australia, the renewal of woodland and forest life hinges on this specific 'Yes!' A new generation of trees is carried on the fur and the tongue, and on the wings that beat through the night carrying the animal to the tree, and carrying the tree's gifts along to other trees. And a new generation of flying-foxes is nurtured into life with lashings of glorious nectar.

There is a foundational ethics within this multispecies kiss. This is the kiss that teaches us that others enable our lives, that others come first because we owe our lives to them. As an ethics, this beautiful, located, embodied, life-affirming, mutually giving and receiving, historical and future-oriented kiss offers a great and seductive lesson: to be for one's self, one must always, also, be for others.

In their refusal to abandon flying-foxes, carers experienced compassion in the face of suffering, and awe in the face of the unknown.

We see this awe in moments of encounter where unknowability becomes a joy. Denise Wade wrote this account:

> Amongst the joys and sadness, trials and tribulations, we are indeed fortunate to share a temporary bond with such sensitive and intelligent mammals and to be privy to some exceptionally beautiful interactions between our two species.
>
> One very special afternoon in February springs to mind that will forever remain in my memory. We had recently received two Little Red flying-foxes into care off separate barbed wire fences . . . One of them had been badly injured on the wire and had required some intensive nursing before being placed into our rehabilitation cage. I was out medicating not long after they arrived and was standing very quietly attempting to coax a rather stubborn patient to take his medication. To my total and utter disbelief, the Little Reds sidled up to me in unison, one on either side and literally wrapped my head in their soft wings whilst gently pressing their warm, furry bodies to my forehead.
>
> I cannot explain the motivation behind this act and nor do I care to but we remained as one for about 10 minutes before they broke contact and gently unfurled their wings from my head. An inexplicable but intensely personal encounter with a beautiful but sadly misunderstood native species.[36]

More becomings

I have thought deeply about waves of ancestral power, and the transition from outside to inside, and yet almost everything remains unknowable. Old Tim Yilngayarri, the clever man, approached the transition zone, but his gift took him just so far. He would only go to the edge of the wave and until it was time for him to cross over, his proper work required him to stop. It was multispecies work that he did, and from there he guided many, many individuals, not only humans but emus who had been wrongly shot, gleaming trees that had been cut down for fence posts, and dingos on the edge of mass extinction.

In this transition zone, individuals cross from outside to inside ancestral waves, and though we know so little, we can try to imagine the quality of the transformation. We think of ancestors as those who have gone before, but the ancestral wave also powers the pulse, sending the pulse from inside to outside.

Waves of powerful, generative becomings open possibilities for the mutualisms and connectivities that bring forth the life that we already know to be beautifully generous. And so, it is brought forth, pulsing into the ephemeral world with iridescent brilliance.

I keep thinking of the tiny psyllids in their protective sugar houses. They aren't annual or predictable. They're not large and showy. Indeed, they are not audible. But their arrival is carried on waves, and when they flourish, everyone benefits.

I think of those pulses, soundless and yet beckoning, bringing news of wild generosity. These silent heralds invite Aboriginal women to harvest and transform the little shelters not just into food but into glistening ceremonial power. Flying-foxes arrive in clouds of enthusiasm, and there will be tongues and blossoms, along with nectar and sugar. There will be colour, light, love and probably wild excitement.

Within the wave there will be those who have witnessed heat stress, torture, mass death and starvation. In this liminal space between horror and celebration, like the beach hibiscus between salt water and fresh, a special power arises.

It is hard to keep my imagination on the inside of this great wave. Beneath the swell of sea: the bright gleam of the shark's eye; against the twilight sky: the clouds of a million flying-foxes; in the opening and closing of the land itself: the news goes forth. Life continues across the shimmering country.

Yes! Becoming ancestral across generations and boundaries. Becoming part of the great shimmer, from outside to inside, and from inside back out again. From being carried to pulsing forth.

Notes

I Speaking of love and peril

1 Leslie Hall and Greg Richards, *Flying Foxes: Fruit and Blossom Bats of Australia* (Sydney: University of New South Wales Press, 2000), 2. Three others are named, but there are no extant colonies, and little is known about them.

2 I explore the connections between ecocide and genocide in Deborah Bird Rose, *Reports from a Wild Country: Ethics for Decolonisation* (Sydney: University of New South Wales Press, 2004).

3 For an example of kinship and nourishment, see Deborah Bird Rose, 'Death and Grief in a World of Kin', in *The Handbook of Contemporary Animism*, ed. Graham Harvey (Durham: Acumen, 2013), 137–47.

4 Also stated as 'ethics as first philosophy'. See, for example, Emmanuel Levinas, *Totality and Infinity: An Essay on Exteriority*, trans. Alphonso Lingis (Pittsburgh, PA: Duquesne University Press, 1969).

5 John Roth, *Ethics after the Holocaust: Perspectives, Critiques, and Responses* (St. Paul, MN: Paragon House, 1999), xvi.

6 Emmanuel Levinas and Richard Kearney, 'Dialogue with Emmanuel Levinas', in *Face to Face with Levinas*, ed. Richard Cohen (Albany: State University of New York Press, 1986).

7   For Levinas and 'Nature' (including animals), see William Edelglass,
     James Hatley, and Christain Diehm, eds, *Facing Nature: Levinas
     and Environmental Thought* (Pittsburgh, PA: Duquesne University
     Press, 2012). On responsiveness to calls of Earth, see James Hatley,
     'The Virtue of Temporal Discernment: Rethinking the Extent and
     Coherence of the Good in a Time of Mass Species Extinction',
     Environmental Philosophy 9, no. 1 (2012).

8   Adam Zachary Newton, *Narrative Ethics* (Cambridge, MA: Harvard
     University Press, 1995), 12.

9   On binaries and hyper-separation, see Val Plumwood, *Environmental
     Culture: The Ecological Crisis of Reason* (London: Routledge, 2002).

10   My translation, apologies for infelicities. Dominique Lestel, *L'Animal
     est l'avenir de l'homme: Munitions pour ceux qui veulent (toujours)
     défendre les animaux* (Paris: Fayard, 2010), 29; Dominique Lestel,
     'La Haine de l'animal', in *Aux origines de l'environnement*, ed. Pierre-
     Henri Gouyon and Hélène Leriche (Paris: Fayard, 2010).

11   Lynn Turner, 'When Species Kiss: Some Recent Correspondence
     between Animots', *Humanimalia* 2, no. 1 (2010), 68.

12   Frans de Waal, *Are We Smart Enough to Know How Smart Animals
     Are?* (New York: W. W. Norton, 2016), 11.

13   On plants, see Peter Wohlleben, *The Hidden Life of Trees*, trans. Jane
     Billinghurst (London: William Collins, 2016); Monica Gagliano, *Thus
     Spoke the Plant: A Remarkable Journey of Groundbreaking Scientific
     Discoveries and Personal Encounters with Plants,* (Berkeley, CA:
     North Atlantic Books, 2018); Lynn Margulis and Dorion Sagan, *What
     Is Life?* (Berkeley: University of California Press, 2000).

14   Donna Haraway, *When Species Meet* (Minneapolis: University of
     Minnesota Press, 2008).

15   Eben Kirksey and Stefan Helmreich, 'The Emergence of Multispecies
     Ethnography', *Cultural Anthropology* 25, no. 4 (2010).

16   Jesper Hoffmeyer, ed., *A Legacy for Living Systems: Gregory Bateson as
     Precursor to Biosemiotics* (New York: Springer, 2008).

17   Brett Buchanan, *Onto-Ethologies: The Animal Environments of
     Uexküll, Heidegger, Merleau-Ponty, and Deleuze* (Albany: State
     University of New York Press, 2008).

18   Marc Bekoff, *The Animal Manifesto: Six Reasons for Expanding Our Compassion Footprint* (Novato, CA: New World Library, 2010).

19   Dominique Lestel, Florence Brunois, and Florence Gaunet, 'Etho-Ethnology and Ethno-Ethology', *Social Science Information* 45, no. 2 (2006), 171.

20   This section owes its origins to Thom van Dooren and Deborah Bird Rose, 'Lively Ethography: Storying Animist Worlds', *Environmental Humanities* 8, no. 1 (2016).

21   Clifford Geertz, 'Ethos, World View, and the Analysis of Sacred Symbols', in *The Interpretation of Cultures* (New York: Basic Books, 1996).

22   Michael Herzfeld, *Anthropology: Theoretical Practice in Culture and Society* (Malden and Oxford: Blackwell Publishers, 2001), 283–4. Not only humans have been known by their ethos. Homer, for example, wrote of the ethos of horses – their habits and habitats.

23   William Dietrich, *The Final Forest: The Battle for the Last Great Trees of the Pacific Northwest* (New York: Penguin Books, 1992), 110.

24   Maria Puig de la Bellacasa, '"Nothing comes without its world": Thinking with Care', *The Sociological Review* 60, no. 2 (2012).

25   Val Plumwood, 'Nature in the Active Voice', *Australian Humanities Review* 46 (2009), 120.

26   See Roth, *Ethics After the Holocaust.*

27   Anna Tsing, 'Arts of Inclusion, or, How to Love a Mushroom', *Australian Humanities Review* 50 (2011), 19.

28   Kirksey and Helmreich, 'The Emergence of Multispecies Ethnography'.

29   Thom van Dooren, Eben Kirksey, and Ursula Münster, 'Multispecies Studies: Cultivating Arts of Attentiveness', *Environmental Humanities* 8, no. 1 (2016), 2.

30   For example, see Plumwood, 'Nature in the Active Voice'.

31   Mary Graham, 'Some Thoughts on the Philosophical Underpinnings of Aboriginal Worldviews', *Australian Humanities Review* 45 (2008), 181.

32   Karen Barad, *Meeting the Universe Halfway: Quantum Physics and the Entanglement of Matter and Meaning* (Durham, NC: Duke University Press, 2007).

33   J. V. Ward and Jack A. Stanford, 'Ecological Connectivity in Alluvial River Ecosystems and its Disruption by Flow Regulation', *Regulated Rivers: Research and Management* 11, no. 1 (1995).

34   On the 'flame of life', see Eileen Crist, 'Intimations of Gaia', in *Gaia in Turmoil: Climate Change, Biodepletion, and Earth Ethics in an Age of Crisis*, ed. Eileen Crist and H. Bruce Rinker (Cambridge, MA: MIT Press, 2010), 328–9.

35   Crist, 'Intimations of Gaia', 329.

36   See Stuart Cooke, 'What are the Animals Saying?' *Plumwood Mountain: An Australian Journal of Ecopoetry and Ecopoetics*, https:// plumwoodmountain.com/what-are-the-animals-saying/

37   Stephan Harding and Lynn Margulis, 'Water Gaia: 3.5 Thousand Million Years of Wetness on Planet Earth', in *Gaia in Turmoil*, ed. Crist and Rinker, 46, 55.

38   Val Plumwood, 'Shadow Places and the Politics of Dwelling', *Australian Humanities Review* 48 (2008).

39   James Hatley, 'The Anarchical Goodness of Creation: Monotheism in Another's Voice', in *Facing Nature*, ed. Edelglass, Hatley, and Diehm.

40   Emmanuel Levinas, *Difficult Freedom: Essays on Judaism*, trans. Sean Hand (Baltimore, MD: Johns Hopkins University Press, 1997).

41   Donna Haraway, 'Situated Knowledges: The Science Question in Feminism and the Privilege of Partial Perspective', *Feminist Studies* 14, no. 3 (1988).

42   Ilya Prigogine, *The End of Certainty: Time, Chaos and the New Laws of Nature* (New York: The Free Press, 1997).

2 Meet the Pteropids

1   Hall and Richards, *Flying Foxes*, 21.

2   Nancy B. Simmons, 'Taking Wing', *Scientific American* 229, no. 6 (2008); Charles Calisher et al., 'Bats: Important Reservoir Hosts of Emerging Viruses', *Clinical Microbiology Reviews* 19, no. 3 (2006), 535.

3   Calisher et al., 'Bats'.

4   Simmons, 'Taking Wing'.

5   Tessa Laird, *Bat* (London: Reaktion Books, 2018).

6   Hall and Richards, *Flying Foxes*, 3.

7   John Pettigrew, 'Are Flying Foxes Really Primates?' *Bats Magazine* 3, no. 2 (Summer 1986).

8   Norberto P. Giannini and Nancy B. Simmons, 'A Phylogeny of Megachiropteran Bats (Mammalia: Chiroptera: Pteropodidae) Based on Direct Optimization Analysis of One Nuclear and Four Mitochondrial Genes', *Cladistics* 19, no. 6 (2003). From Melanesia and Southeast Asia, megabats radiated both east and west, colonising Africa several times. The megachiropterans are known as the Pteropodidae family; I will not use this latter term in order to avoid possible confusion.

9   Mary White, *Running Down: Water in a Changing Land* (Sydney: Kangaroo Press, 2000), 2.

10 Keith F. Walker, Jim T. Puckridge, and Stuart J. Blanch, 'Irrigation Development on Cooper Creek, Central Australia: Prospects for a Regulated Economy in a Boom-and-Bust Ecology', *Aquatic Conservation Marine and Freshwater Ecosystems* 7, no. 1 (1997).

11 White, *Running Down*, 1–2.

12 Anna Vidot, 'Flying foxes crossing Bass Strait', ABC Rural, 30 August 2010, http://www.abc.net.au/site-archive/rural/tas/content/2010/08/s2997424.htm

13 Hall and Richards, *Flying Foxes*, 1–3.

14 Hall and Richards, *Flying Foxes*, 20.

15 Interview with Tim Pearson, 8 June 2010.

16 Justin Welbergen and Peggy Eby, 'Not in my backyard? How to live alongside flying-foxes in urban Australia', *The Conversation*, 27 May, 2016, https://theconversation.com/not-in-my-backyard-how-to-live-alongside-flying-foxes-in-urban-australia-59893. On federal monitoring, see 'Monitoring Flying-Fox Populations', Australian Government: Department of the Environment and Energy, https://www.environment.gov.au/biodiversity/threatened/species/flying-fox-monitoring

17 Natasha Fijn, 'Following Flying Foxes III: What happens when humans decide on zero tolerance?' 2011, https://vimeo.com/30881933

18 Fijn, 'Following Flying Foxes III'.

19   Hall and Richards, *Flying Foxes*, 24–8.

20   Hall and Richards, *Flying Foxes*, 42–6.

21   Hall and Richards, *Flying Foxes*, 46.

22   Hall and Richards, *Flying Foxes*, 64.

23   Interview with Tim Pearson, 8 June 2010. When I checked in with Tim a few years later he informed me that the bats in Gordon have now moved closer to houses and the uneasy truce has collapsed.

24   Fijn, 'Following Flying Foxes III'.

25   Interviews with Tim Pearson, 8 June 2010 and 15 June 2010.

26   Interviews with Pearson.

27   K. A. Parry-Jones and M. L. Augee, 'Movements of Grey-Headed Flying Foxes (*Pteropus poliocephalus*) to and from a Colony Site on the Central Coast of New South Wales', *Wildlife Research* 19 (1992), 337.

28   In the mid-1990s the Eucalyptus family was subdivided into Eucalypts and Corymbia. In keeping with common practice, I use the term Eucalypt to include the Corymbias.

29   Peggy Eby, ed., *The Biology and Management of Flying Foxes in NSW* (Sydney: NSW National Parks and Wildlife Service, 1995), 35–6.

30   Parry-Jones and Augee, 'Movements of Grey-Headed Flying Foxes'.

31   Eby, ed., *The Biology and Management*, 36–7.

32   Eby, ed., *The Biology and Management*, 37–8.

33   Hall and Richards, *Flying Foxes*, 69.

34   Hall and Richards, *Flying Foxes*, 78.

35   Interview with Tim Pearson, 8 June 2010.

36   Hall and Richards, *Flying Foxes*, 82.

37   Hall and Richards, *Flying Foxes*, 80–3.

38   Carol Booth et al., *Why NSW Should Ban the Shooting of Flying Foxes* (Sydney: Humane Society International, 2008), 5–6, 12.

39   Tim Low, *Where Song Began: Australia's Birds and How They Changed the World* (Sydney: Penguin Random House Australia, 2014), 16.

40   K. Parry-Jones and M. L. Augee, 'Food Selection by Grey-Headed Flying Foxes (*Pteropus Poliocephalus*) Occupying a Summer Colony Site near Gosford, New South Wales', *Wildlife Research* 18, no. 1 (1991), 111.

41 Christopher R. Tidemann and Michael J. Vardon, 'Pests, Pestilence, Pollen and Pot Roasts: The Need for Community Based Management of Flying Foxes in Australia', *Australian Biologist* 10, no. 1 (March 1997), 79–80.

42 Eby, ed., *The Biology and Management*, 25.

43 Parry-Jones and Augee, 'Food Selection', 122.

44 Nicola Markus and Les Hall, 'Foraging Behaviour of the Black Flying-Fox (*Pteropus Alecto*) in the Urban Landscape of Brisbane, Queensland', *Wildlife Research* 1, no. 3 (2004), 346.

45 Markus and Hall, 'Foraging Behaviour'.

46 Interview with Tim Pearson, 8 June 2010.

47 Interview with Pearson. Since this interview was conducted the shooting of flying-foxes in NSW has declined. According to Tim this 'has been one of our rare wins, [a result of] the combination of subsidised netting, a more difficult commercial situation, and many small peri-urban orchards being more profitably sold for housing development'. Both legal and illegal shooting are now rare in NSW. Queensland, however, is a different story.

48 Interview with Pearson.

49 Personal communication with Mick Bower, a fine gentleman of the old bush ways, and a keen observer of life in his town of Katherine (NT).

50 Hall and Richards, *Flying Foxes*, 50–4.

51 For a fuller discussion of these diseases, see Chapter 7. Be warned that one should never handle flying-foxes without having been vaccinated. If bitten or scratched, a person should seek immediate expert care. Early follow-up treatment prevents the virus from developing.

52 Melanie S. Thomson, 'Placing the Wild in the City: "Thinking With" Melbourne's Bats', *Society and Animals* 15, no. 1 (2007), 79–95; Tim Low, *The New Nature* (Melbourne: Viking, 2002), 47.

53 James Woodford, 'The swingers', *The Sydney Morning Herald*, 23 April 2003.

54 Dr David Westcott, CSIRO, Atherton stated in a recent public forum that the number of camps inhabited by spectacled flying-foxes is decreasing, while the numbers of animals per camp is increasing, and that the greatest increase is in urban camps. He identified urbanisation

as a process that is taking place amongst numerous species, and he further stated that scientists do not at this time have an explanation for the process.

55 Kristen Parris and D. L. Hazell, 'Biotic Effects of Climate Change in Urban Environments: The Case of the Grey-Headed Flying-Fox (*Pteropus poliocephalus*) in Melbourne, Australia', *Biological Conservation* 124 (2005).

56 Thomson, 'Placing the Wild in the City'.

57 Low, *The New Nature*, 309.

58 Dan Perry, '"Endemic Aliens": Grey-Headed Flying-Foxes at the Melbourne Royal Botanic Gardens', *Journal of Animal Ethics* 2, no. 2 (2012), 174.

59 Interview with Tim Pearson, 8 June 2010.

60 Interview with Jenny Mclean, 14 August 2010.

61 Many volunteers helped in this project, which was coordinated by Lawrence Pope and the organisation Victorian Advocates of Animals.

62 Low, *The New Nature*, 11.

63 Francis Ratcliffe, *Flying Fox and Drifting Sand: The Adventures of a Biologist in Australia* (Sydney: Angus and Robertson, 1948), 120.

64 Dr Rodney van der Ree, Royal Botanic Gardens, Melbourne. Dr van der Ree was in charge of the operation. He offered this figure in the course of a public speech in Cairns, 13 August 2010.

65 To see a bit of the video, visit Deborah Bird Rose, 'Flying-foxes at Port Keats 2013', https://vimeo.com/75266158

3 Arts of care

1 Peter Hannam, 'Records melt in our hottest year', *The Sydney Morning Herald*, 21 December 2013, https://www.smh.com.au/environment/climate-change/records-melt-in-our-hottest-year-20131220-2zqrt.html

2 Malcolm Holland, '15,000 flying foxes killed in heat in Sydney', *The Daily Telegraph*, 23 January 2013, https://www.dailytelegraph.com.au/flying-foxes-killed-in-heat-in-sydney/news-story/86dc7bfeb25ee5cb0611e6534e5c9d62

3   Low, *Where Song Began*, 54.
4   Josh Bavas, 'About 100,000 bats dead after heatwave in Southern Queensland', ABC, 8 January 2014, http://www.abc.net.au/news/2014-01-08/hundred-thousand-dead-bats-after-qld-heatwave-rspca-says/5190644
5   This phrase is widely quoted; see, for example, Tony Mohr, 'Weather on Steroids: Climate Change in Action', ABC, 7 January 2013, http://www.abc.net.au/news/2013-01-07/mohr-weather-on-steroids/4455022
6   Bavas, 'About 100,000 Bats Dead'.
7   Justin Welbergen, Carol Booth, and John Martin, 'Killer climate: Tens of thousands of flying foxes dead in a day', *The Conversation*, 25 February 2014, https://theconversation.com/killer-climate-tens-of-thousands-of-flying-foxes-dead-in-a-day-23227. The heatwave continued, and affected a much larger area, so it is reasonable to extrapolate to larger figures.
8   Interview with Tim Pearson, 8 June 2010.
9   Interview with Denise Wade, 13 August 2010; other quotes are from the same interview.
10 Regent's Park is a suburban area south of Brisbane.
11 L. Martin and A. P. McIlwee, 'The Reproductive Biology and Intrinsic Capacity for Increase of the Grey-Headed Flying-Fox Poliocephalus (Megachiroptera), and the Implications of Culling', in *Managing the Grey-Headed Flying-Fox as a Threatened Species in NSW*, ed. Peggy Eby and Dan Lunney (Sydney: Royal Zoological Society of New South Wales, 2002). Statements about flying-fox behaviour in the bush are based on observational evidence and, given the size of camps and the numbers of individuals involved, it must be said that generalisations about adoption should be taken as somewhat provisional.
12 Unpublished correspondence to members of Bat Conservation & Rescue QLD (BCRQ). Reproduced with permission of the author.
13 Emmanuel Levinas, 'The Paradox of Morality: An Interview with Emmanuel Levinas', in *The Provocation of Levinas: Rethinking the Other*, ed. Robert Bernasconi and David Woods (London: Routledge, 1988), 163.
14 His optimism towards future deathwork was not misplaced: 'Each

year, the department issues a limited number of damage mitigation permits (DMPs) under the Act allowing the "lethal take" of flying-foxes by shooting in order to protect commercial fruit crops.' See 'Damage mitigation permits for crop protection', Queensland Government: Environment, https://www.ehp.qld.gov.au/wildlife/liv ingwith/flyingfoxes/damage-mitigation-permits.html

15   Taken from 'Bat Conservation & Rescue QLD, Inc', www.bats.org.au. Reproduced with permission.

16   Hall and Richards, *Flying Foxes*, 42.

17   Interview with Jenny Mclean, 12 August 2010.

18   Interview with Tim Pearson, 8 June 2010.

19   Interview with Denise Wade, 13 August 2010.

20   Puig de la Bellacasa, "'Nothing comes without its world'".

21   Thom van Dooren, 'Care', *Environmental Humanities* 5, no. 1 (2014).

22   Puig de la Bellacasa, "'Nothing comes without its world'", 200–1.

23   Puig de la Bellacasa, "'Nothing comes without its world'", 201.

24   I undertook training with an organisation called Sydney Wildlife and thus became familiar with their guidelines.

25   Frederique Apffel-Marglin and Stephen Marglin, *Decolonizing Knowledge: From Development to Dialogue* (Oxford: Clarendon Press, 1996). A related term is the Greek word '*metis*'. Scholars differ in the exact definitions of these terms; see, for example, James C. Scott, *Seeing Like a State: How Certain Schemes to Improve the Human Condition Have Failed* (New Haven, CT: Yale University Press, 1998).

26   Wohlleben, *The Hidden Life of Trees*.

27   Frans de Waal, 'Putting the Altruism Back into Altruism: The Evolution of Empathy', *Annual Review of Psychology* 59 (2008).

28   Dominique Lestel, 'The Question of the Animal Subject: Thoughts on the Fourth Wound to Human Narcissism', trans. Hollis Taylor, *Angelaki* 19, no. 3 (2014) (emphasis in the original).

29   Thom van Dooren, *The Wake of Crows: Living and Dying in Shared Worlds* (New York: Columbia University Press, 2019), 157.

30   Val Plumwood, *Feminism and the Mastery of Nature* (London: Routledge, 1993), 213.

31   Interview with Louise Saunders, 12 August 2014.

32   Interview with Tim Pearson, 8 June 2010. In addition to his volunteer work as carer, researcher, advocate and public educator, Tim carried out doctoral research into flying-fox communication.

33   See Natasha Fijn, 'Following Flying Foxes II: What is it like to foster an orphan?', 2011, https://vimeo.com/30877202

34   Fijn, 'Following Flying Foxes II'.

35   Hall and Richards, *Flying Foxes*, 45.

36   See 'Bat wraps #3: Giving injured flying foxes a second chance', 2011, https://www.youtube.com/watch?v=mDEx0sLgoHk. The video is introduced by the noted scientist, educator, mediator and advocate Peggy Eby.

37   Elizabeth Larsen et al., 'Neighbours of Ku-ring-gai Flying-fox Reserve: Community Attitudes Survey 2001', in *Managing the Grey-Headed Flying-Fox as a Threatened Species in New South Wales*, ed. Peggy Eby and Daniel Lunney (Sydney: Royal Zoological Society of New South Wales, 2002).

38   Interview with Tim Pearson, 8 June 2010.

39   Levinas and Kearney, 'Dialogue with Emmanuel Levinas', 23–4.

40   Judith Butler, *Precarious Life: The Powers of Mourning and Violence* (London: Verso, 2004), 134.

41   Dominique Lestel, 'Eprouver la personne comme personnage', in *Personne/Personnage*, ed. Thierry Lenain and Aline Wiame (Paris: Vrin, 2011).

42   See Edelglass, Hatley, and Diehm, eds, *Facing Nature*, in particular Hatley, 'The Anarchical Goodness of Creation'.

43   Mark Rowlands, *Can Animals Be Moral?* (Oxford: Oxford University Press, 2012), 71–98.

44   Rowlands, *Can Animals Be Moral?*, 197.

45   Interview with Louise Saunders, 12 August 2014. One might well consider Derrida's famous question in light of flying-fox communicative reciprocity: 'And say the animal responded?' See Jacques Derrida, *The Animal That Therefore I Am*, ed. Marie-Louise Mallet, trans. David Wills (New York: Fordham University Press, 2008).

4 Participation

1   Lucien Lévy-Bruhl, *Primitive Mythology: The Mythic World of the Australian and Papuan Natives*, trans. Brian Elliott (St Lucia: Queensland University Press, 1983), 23. He states, 'their metaphysics is quite spontaneous; it is the result of the frequent, one might say constant, experience of a reality which goes beyond and dominates all common nature, and yet is present and active in it at all times'.

2   Stanley Tambiah, *Magic, Science, Religion and the Scope of Rationality* (Cambridge: Cambridge University Press, 1990), 86–7.

3   Lucien Lévy-Bruhl, *How Natives Think*, trans. Lilian Clare (London: George Allen & Unwin, 1926), 69–104.

4   David Abram, *The Spell of the Sensuous: Perception and Language in a More-Than-Human World* (New York: Vintage Books, 1996), 57.

5   As quoted in Lev Shestov, 'Myth and Truth: On the Metaphysics of Knowledge', in *Speculation and Revelation* (Athens: Ohio University Press, 1982), 124–5.

6   His work is thus delightfully consistent with contemporary thinking in biology, biosemiotics, systems theory and ecology.

7   Such control was labelled 'protection'.

8   The best account of the decades of cattle station life, when people were 'prisoners in their own country', is offered in Hobbles Danaiyarri, 'The Saga of Captain Cook', in *Australia's Empire*, ed. Deryck Schreuder and Stuart Ward (Oxford: Oxford University Press, 2008); Deborah Bird Rose, 'The Saga of Captain Cook: Morality in Aboriginal and European Law', *Australian Aboriginal Studies* 2 (1984).

9   See Donald Grinde and Bruce Johansen, *Ecocide of Native America: Environmental Destruction of Indian Lands and Peoples* (Santa Fe, NM: Clear Light Publishers, 1995).

10 See Deborah Bird Rose, *Hidden Histories: Black Stories from Victoria River Downs, Humbert River, and Wave Hill Stations* (Canberra: Aboriginal Studies Press, 1991), 245.

11 As I gained experience in Land Rights hearings, I worked on about eighteen land claims (including disputes) across the NT from the deserts to the Gulf of Carpentaria. I worked with Yarralin people

on their claims to land, and on registering sacred sites. Some of these claims took years to resolve, but the legislation came with a sunset clause, and that phase of restorative justice is now finished. See Deborah Bird Rose, 'Country for Yarralin', 7 July 2016, https://webarchive.nla.gov.au/tep/177305

12   Between 1980 and 2018 the percentage of whitefellas jumped from .025 per cent to 8.6 per cent. Amenities increased, and so too, in keeping with government policy, did the scrutiny, oversight and control. For a discussion of government policy, see Deborah Bird Rose, 'Remembrance, in the Wake of Suicide', 28 March 2016, https://webarchive.nla.gov.au/tep/177305

13   I have written extensively on these matters in other publications. See Rose, *Hidden Histories*; Deborah Bird Rose, *Dingo Makes Us Human: Life and Land in an Australian Aboriginal Culture* (Cambridge: Cambridge University Press, 1992); Rose, *Reports from a Wild Country*; Deborah Bird Rose, *Wild Dog Dreaming: Love and Extinction* (Charlottesville: University of Virginia Press, 2011).

14   Graham Harvey, *Animism: Respecting the Living World* (New York: Columbia University Press, 2006), xi. See also Harvey, ed., *The Handbook of Contemporary Animism*.

15   The term originates in the Ojibwa language of North America.

16   Fiona Magowan, *Melodies of Mourning: Music and Emotion in Northern Australia* (Nedlands: University of Western Australia Press, 2007). She uses the term 'mutual indwelling' to identify some of the deep implications of co-substantiality.

17   A. P. Elkin, *The Australian Aborigines: How to Understand Them* (Sydney: Angus and Robertson, 1954), 133.

18   In classic anthropology the emphasis is primarily on biological descent. More nuanced approaches which emphasise kinship as an outcome of nurturing relationships provide a balance to the earlier emphasis on descent.

19   Kinship is one of the perennial topics in the disciplines of anthropology. On the matter of considering both practice and structures, see, for example, David Schneider, *A Critique of the Study of Kinship* (Ann Arbor: University of Michigan Press, 1984); also Maximilian Holland,

Social Bonding and Nurture Kinship: Compatibility between Cultural and Biological Approaches (North Charleston, SC: CreateSpace Press, 2012).

20 This is not a complete analysis of the complexities. For greater detail, see Rose, *Dingo Makes Us Human*.

21 This story is recounted in greater detail in Rose, *Dingo Makes Us Human*, 83–5.

22 Deborah Bird Rose, *Nourishing Terrains: Australian Aboriginal Views of Landscape and Wilderness* (Canberra: Australian Heritage Commission, 1996).

23 For an excellent analysis, see Elizabeth Povinelli, *Labor's Lot: The Power, History, and Culture of Aboriginal Action* (Chicago: University of Chicago Press, 1993).

24 Catherine Ellis, *Aboriginal Music, Education for Living: Cross Cultural Experiences from South Australia* (St Lucia: University of Queensland Press, 1985), 82–3.

25 Ellis, *Aboriginal Music*, 92–3. Ellis notes that the pattern of the music is retained perfectly across the breaks within which there is no music.

26 Hall and Richards, *Flying Foxes*, 53.

27 This account of the women was provided by a senior Lawwoman. There is no authorised version, and inevitably different people emphasise different parts of the story.

28 All the information presented here is in the public realm and has been discussed and documented in the course of various claims to land.

29 See David McKnight, 'Men, Women, and Other Animals: Taboo and Purification among the Wikmungkan', in *The Interpretation of Symbolism*, ed. Roy Willis (London: Malaby Press, 1975).

30 I was unable to obtain a reliable identification. All or most of the microbats are grouped together under a single term: *pangkal*.

31 Administered by the Aboriginal Areas Protection Authority.

32 A more detailed account was published in Deborah Bird Rose, 'On History, Trees and Ethical Proximity', *Postcolonial Studies* 11, no. 2 (2008).

33 As stated in Chapter 2, I use the term Eucalypt to include the Corymbias.

34   Although I use the conventional term 'painting', Yarralin people in general hold these images to be actual instantiations of the beings that, from a western point of view, they represent. See Darrell Lewis and Deborah Bird Rose, *The Shape of the Dreaming: The Cultural Significance of Victoria River Rock Art* (Canberra: Aboriginal Studies Press, 1988).

35   See A. R. Radcliffe-Brown, 'The Rainbow-Serpent Myth in South-East Australia', *Oceania* 1, no. 3 (1930).

36   This particular shrub was *Terminalia erythrocarpa*. In nearby areas people say the flying-foxes eat the fruits; Daly and I were focused on leaves and may have neglected the more obvious matter of the fruits.

37   For a somewhat sensationalised video of flying-foxes and crocodiles in a northern tropical river, see 'Flying Foxes vs. Freshwater Crocodile | Lands of the Monsoon | BBC Earth', BBC Earth, 10 April 2015, https://www.youtube.com/watch?v=wi30w-Mk2yQ

5 Nomads

1   Eby, *The Biology and Management*. Until recently the terms nomadic and migratory were used interchangeably but it is now established practice to distinguish the two and to class flying-foxes as nomadic. The importance of nomadism to information exchange was discussed by Peggy Eby in a public speech she gave on 13 August 2010.

2   Hall and Richards, *Flying Foxes*.

3   Billie J. Roberts, Carla P. Catterall, Peggy Eby, and John Kanowski, 'Long-Distance and Frequent Movements of the Flying-Fox *Pteropus poliocephalus*: Implications for Management', *PLoS ONE* 7, no. 8 (2012).

4   Christopher R. Tidemann and J. E. Nelson, 'Long-Distance Movements of the Grey-Headed Flying Fox (*Pteropus Poliocephalus*)', *Journal of Zoology* 263, no. 2 (2004), 143.

5   Tidemann and Nelson, 'Long-Distance Movements', 144.

6   Welbergen and Eby, 'Not in my backyard?'

7   The story was reported in numerous media; see, for example, 'Flying

Fox Dispersal: Fact Sheet', Eurobodalla Shire Council, http://
www.riverbendnelligen.com/downloads/flyingfox2.pdf; 'NSW govt
"destruction of flying fox habitat"', Echo Netdaily, 26 May 2016,
https://www.echo.net.au/2016/05/nsw-govt-approves-destruction-of-
flying-fox-habitat/

8   'Flying-foxes', Eurobodalla Shire Council, http://www.esc.nsw.gov.au/
living-in/about/our-natural-environment/grey-headed-flying-foxes

9   H. J. Spencer, C. Palmer, and K. Parry-Jones, 'Movements of Fruit-
Bats in Eastern Australia, Determined by Using Radio-Tracking',
Wildlife Research 18, no. 4 (1991), 46.

10   Hall and Richards, *Flying Foxes*, 46.

11   Zhanna Reznikova, *Animal Intelligence: From Individual to Social
Cognition* (Cambridge: Cambridge University Press, 2007), 104.

12   John Edward Huth, *The Lost Art of Finding Our Way* (Cambridge,
MA: The Belknap Press of Harvard University Press, 2013), 23.

13   Reznikova, *Animal Intelligence*, 108.

14   Huth, *The Lost Art*, 28.

15   Reznikova, *Animal Intelligence*, 120–1.

16   Birds have their own equivalent to the hippocampus and many navigate by cognitive maps. Bird forms of navigation are extremely varied
and complex and are beyond the scope of my discussion.

17   Huth, *The Lost Art*, 26–8.

18   Reznikova, *Animal Intelligence*, 105.

19   Joan Didion, *The White Album* (New York: Farrar, Strauss & Giroux,
1990). The terms 'narrative' and 'story' can be defined in ways that
separate them, but here I use them interchangeably.

20   I avoid the term roost fidelity; another technical term is philopatry.
For the sake of clarity, I stick to one term only.

21   Jesper Hoffmeyer, 'From Thing to Relation: On Bateson's
Bioanthropology', in *A Legacy for Living Systems*, ed. Hoffmeyer,
38.

22   Paul Shepard, *The Others: How Animals Made Us Human* (Washington,
DC: Island Press, 1996), 16–18, 21. Some of the information in this
section was previously published; see Thom van Dooren and Deborah
Bird Rose, 'Storied-Places in a Multispecies City', *Humanimalia* 3,

no. 2 (Spring 2012). On other animals, see Eileen Crist, *Images of Animals: Anthropomorphism and Animal Mind* (Philadelphia, PA: Temple University Press, 1999), 170–1.

23   Hall and Richards, *Flying Foxes*, 94.

24   Eby, *The Biology and Management*, 18.

25   Peggy Eby, 'Seasonal Movements of Grey-Headed Flying-Foxes, *Pteropus Poliocephalus* (Chiroptera: Pteropodidae) from Two Maternity Camps in Northern New South Wales', *Wildlife Research* 18 (1991), 549.

26   L. A. Shilton et al., 'Landscape-Scale Redistribution of a Highly Mobile Threatened Species, *Pteropus Conspicillatus* (Chiroptera, Pteropodidae), in Response to Tropical Cyclone Larry', *Austral Ecology* 33, no. 4 (2008). There was no suggestion that the missing 50,000 individuals had died. Rather it was presumed that they had joined other camps.

27   M. J. Vardon et al., 'Seasonal Habitat Use by Flying-Foxes, *Pteropus Alecto* and *P. Scapulatus* (Megachiroptera), in Monsoonal Australia', *Journal of Zoology* 253, no. 4 (2001). This study attempted to correlate *P. alecto* behaviour with the seasons as they are defined by the Gundyemi people of western Arnhem Land.

28   Summarised from Hall and Richards, *Flying Foxes*, 45–6.

29   Interview with Marjorie Beck, 2 June 2010.

30   Hall and Richards, *Flying Foxes*, 94.

31   Interview with Tim Pearson, 8 June 2010.

32   Jennifer Parsons, Simon Robson, and Louise Shilton, 'Roost Fidelity in Spectacled Flying-Foxes *Pteropus Conspicullatus*: Implications for Conservation and Management', in *The Biology and Conservation of Australasian Bats*, ed. Bradley Law et al. (Sydney: Royal Zoological Society of New South Wales, 2011).

33   Interview with Louise Saunders, 12 August 2010. Following quotes from Louise Saunders are from this same interview.

34   James Scott's deep research in Southeast Asia tells us that through time a majority of people were nomadic in one way or another as they sought to escape the horrors of living under oppressive and ruthless state systems. See James Scott, *The Art of Not Being Governed: An*

Anarchist History of Upland Southeast Asia (New Haven, CT: Yale University Press, 2010).

35   For a contemporary Indigenous rejection of nomadism, see Bruce Pascoe, *Dark Emu* (Broome: Magabala Books, 2014). Pascoe is quite correct in noting the process of intensification that was underway in parts of Victoria.

36   Gilles Deleuze is a key figure here; I draw more specifically on the work of Isabelle Stengers. She is deeply influenced by Deleuze, among others; I find that she works interactively with reality in all its impressive complexity and weirdness, since her aim is not, in the end, philosophy, but rather transformation of the current power structures that are destroying Earth.

37   The contrast with contemporary corporate modes of mobility, relying as they do on practices of 'deplete, destroy and depart', could not be more extreme. See Tom Athanasiou, *Slow Reckoning* (Boston, MA: Little, Brown & Co., 1996), 242.

38   D. M. J. S. Bowman, 'Tansley Review No. 101: The Impact of Aboriginal Landscape Burning on the Australian Biota', *New Phytologist* 140 (1998).

39   Deborah Bird Rose et al., *Country of the Heart: An Indigenous Australian Homeland* (Canberra: Aboriginal Studies Press, 2002), 19.

40   My knowledge of the use of fires is drawn from my research with Mak Mak people whose home country is in the floodplains south-west of Darwin. See Rose et al., *Country of the Heart*. One of the first things invading pastoralists did in the savannah regions was to brutally suppress Indigenous fire regimes.

41   John N. Thompson, *The Geographic Mosaic of Coevolution* (Chicago: University of Chicago Press, 2005), 3.

42   Thompson, *Geographic Mosaic*, 3.

43   Lynn Margulis, *Symbiotic Planet: A New Look at Evolution* (Amherst, MA: Basic Books, 1998), 5.

44   Margulis, *Symbiotic Planet*, 247.

45   Margulis, *Symbiotic Planet*, 5.

46   Hall and Richards, *Flying Foxes*, 82–4.

47   The predator-prey relationship can be understood in terms of mutu-
      alisms, but my focus is not on making that case at this time.

48   Isabelle Stengers, *Cosmopolitics I*, trans. Robert Bononno
      (Minneapolis: University of Minnesota Press, 2010), 36.

49   Stengers, *Cosmopolitics I*, 36.

50   Low, *Where Song Began*, 41. Unlike aphids they are not managed by
      other insects.

51   There are two other types of exudates, but I'll stick with psyllids.

52   Low, *Where Song Began*, 41.

53   Low, *Where Song Began*, 34.

54   The category is mangari – plant food, women's food. The other cat-
      egory is ngarin – meat food, men's food. Of course, women hunt
      opportunistically, and men forage opportunistically, but the greatest
      contributions to diet – plant foods and animal foods – are the respon-
      sibility of women and men, respectively.

55   For more analysis of this speculative hypothesis, see my article and
      those of others in a special issue of the journal *Material Religion*,
      Deborah Bird Rose, 'Gendered Substance and Objects in Ritual: An
      Australian Study', *Material Religion: The Journal of Objects, Art and
      Belief* 3, no. 1 (2007).

56   Bowman, 'The Impact of Aboriginal Landscape Burning'; for a
      continent-wide survey, see Bill Gammage, *The Biggest Estate on Earth*
      (Sydney: Allen & Unwin, 2012).

57   Darrell Lewis, *Slower Than the Eye Can See* (Darwin: Tropical
      Savannas CRC, 2002). This book documents the spread of 'woody
      weeds' across the savannah in the Northern Territory.

58   Freya Mathews, 'Earth as Ethic', in *Manifesto for Living in the
      Anthropocene*, ed. Katherine Gibson, Deborah Bird Rose, and Ruth
      Fincher (Brooklyn: Punctum, 2015).

6 Ancestral power

1   Thomas Nagel, 'What Is It Like to Be a Bat?', *The Philosophical Review*
     83, no. 4 (1974). This is the classic essay on the impermeability of
     borders between minds. Over the years research has suggested that

the border between minds is more porous than Nagel would seem to allow.

2   Lev Shestov, *Athens and Jerusalem*, trans. Bernard Martin (New York: Simon and Schuster, 1968), 406–7.

3   Lev Shestov, 'Children and Stepchildren of Time: Spinoza in History', in *A Shestov Anthology*, ed. Bernard Martin (Athens: Ohio University Press, 1970).

4   Shestov, *Athens and Jerusalem*, 378–9.

5   Lev Shestov, 'Speculation and Apocalypse: The Religious Philosophy of Vladimir Solovyov', in *Speculation and Revelation* (Athens: Ohio University Press, 1982), 85.

6   See, for example, Prigogine, *The End of Certainty*.

7   Plumwood, *Feminism and the Mastery of Nature*, 102.

8   Danaiyarri, 'Saga of Captain Cook'. See Chapter 4.

9   Quoted elsewhere. See Rose, *Dingo Makes Us Human*, 57.

10   On these logics, see Chapter 1.

11   Dorion Sagan, 'Life on a Margulisian Planet: A Son's Philosophical Reflections', in *Earth, Life, and System: Evolution and Ecology on a Gaian Planet*, ed. Bruce Clarke (New York: Fordham University Press, 2015), 19.

12   Harding and Margulis, 'Water Gaia', 46, 55.

13   Crist, 'Intimations of Gaia', 328–9. There are numerous analyses of animist thought in relation to the West's old and new sciences. See Plumwood, 'Nature in the Active Voice'; Deborah Bird Rose, 'Val Plumwood's Philosophical Animism', *Environmental Humanities* 3, no. 1 (2013); Harvey, *The Handbook of Contemporary Animism*.

14   Stengers, *Cosmopolitics I*, 36.

15   Stengers, *Cosmopolitics I*, 36.

16   Margulis and Sagan, *What Is Life?*, 91.

17   Hans Jonas, 'The Burden and Blessing of Mortality', *Hastings Center Report* 22, no. 1 (1992), 38. He is writing about humans here, but the argument is applicable across all who arrive in the world newly formed whether by birthing, hatching, sprouting, sporing.

18   Jonas, 'Burden and Blessing', 34.

19   Jonas, 'Burden and Blessing', 36.

20 Quoted in Lev Shestov, 'Myth and Truth', 124–5.

21 Karen Barad, 'Posthumanist Performativity: Toward an Understanding of How Matter Comes to Matter', *Signs* 28, no. 3 (2003), 801–31. I am simplifying Barad's argument without, I hope, losing its force.

22 Barad, 'Posthumanist Performativity', 818.

23 Haraway, *When Species Meet*. See also Thom van Dooren, *Flight Ways: Life and Loss at the Edge of Extinction* (New York: Columbia University Press, 2014), 10.

24 de Waal, *Are We Smart Enough to Know How Smart Animals Are?*, xx; Wohlleben, *The Hidden Life of Trees*, 32–4.

25 Peter Harries-Jones, *A Recursive Vision: Ecological Understanding and Gregory Bateson* (Toronto: University of Toronto Press, 1995), 107.

26 van Dooren, *Flight Ways*, 27, 29.

27 Harding and Margulis, 'Water Gaia', 46, 55.

28 van Dooren, *Flight Ways*, 12.

29 Hatley, 'The Anarchical Goodness of Creation'.

30 On 'man-made mass death' also see Edith Wyschogrod, *Spirit in Ashes: Hegel, Heidegger, and Man-Made Mass Death* (New Haven, CT: Yale University Press, 1985).

31 James Hatley, *Suffering Witness: The Quandary of Responsibility after the Irreparable* (Albany: State University of New York Press, 2000), 61.

32 Hatley, *Suffering Witness*, 60–1.

33 Freya Mathews, *The Ecological Self* (London: Routledge, 1991), 98.

34 Mathews, 'Earth as Ethic', 93.

35 Scott Richard Shaw, *Planet of the Bugs* (Chicago: University of Chicago Press, 2014), 156–7.

36 Matthew Hall, *Plants as Persons: A Philosophical Botany* (Albany: State University of New York Press, 2011), 22–8. New research into chemical communication, and into the underground lives of plants, is rapidly expanding our understanding of the complexity of their lives. See, for example, Wohlleben, *The Hidden Life of Trees*.

37 Botanists refer to lures and rewards to identify these desires and seductions.

38 Martin Burd, 'Colourful language – it's how Aussie birds and flowers "speak"', *The Conversation*, 26 February 2014, https://theconversa tion.com/colourful-language-its-how-aussie-birds-and-flowers-speak-23659

39 Freya Mathews, 'From Biodiversity-based Conservation to an Ethic of Bio-proportionality', *Biological Conservation* 200 (2016).

40 Sean Carroll, *Endless Forms Most Beautiful: The New Science of Evo Devo and the Making of the Animal Kingdom* (London: Quercus, 2011).

41 Howard Morphy, 'From Dull to Brilliant: The Aesthetics of Spiritual Power among the Yolngu', *Man*, New Series, 24, no. 1 (1989).

42 Morphy, 'From Dull to Brilliant', 21.

43 Morphy, 'From Dull to Brilliant', 30.

44 Participation is not always positive; catastrophic events are part of the flow of life, as we have seen.

45 For more detail on women in land claims, see: Deborah Bird Rose, 'Histories and Rituals: Land Claims in the Territory', in *In the Age of Mabo: History, Aborigines and Australia*, ed. Bain Attwood (Sydney: Allen & Unwin, 1996); Deborah Bird Rose, 'Conflict Resolution and Decolonisation: Aboriginal Australian Case Studies in "Enlarged Thinking"', in *Mediating across Difference: Oceanic and Asian Approaches to Conflict Resolution*, ed. Morgan Brigg and Roland Bleiker (Honolulu: University of Hawaii Press, 2011), 100–14.

7 The vortex

1 Shaw, *Planet of the Bugs*, 168–9.

2 I am using the term species for convenience and because it usually does map well onto real differences. It is the subject of debate, and clearly the boundaries between species, especially closely related ones, may be somewhat arbitrary. But in one's everyday experience of the world, different kinds of living beings are observable, and those differences show a regularity and predictability from generation to generation, leading to the understanding that such species' differences are neither random nor solely the result of human efforts to classify the world.

3   Shaw, *Planet of the Bugs*, 181.

4   Edward O. Wilson, *The Future of Life* (New York: Alfred A. Knopf, 2002), 329.

5   Stephan Harding, 'Gaia and Biodiversity', in *Gaia in Turmoil*, ed. Crist and Rinker, 107.

6   Dean Janzen, 'The Deflowering of Central America', *Natural History* 83, no. 4 (1974).

7   W. J. Bond, 'Assessing the Risk of Plant Extinction Due to Pollinator and Disperser Failure', in *Extinction Rates*, ed. John Lawton and Robert May (Oxford: Oxford University Press, 1995), 45.

8   The slogan is a clever reworking of a plea from koalas: No Tree, No Me.

9   Booth et al., *Why NSW Should Ban*, 12.

10   Deborah Bird Rose, Thom van Dooren, and Matthew Chrulew, 'Introduction: Telling Extinction Stories', in *Extinction Studies: Stories of Death, Time and Generations*, ed. Deborah Bird Rose, Thom van Dooren, and Matthew Chrulew (New York: Columbia University Press, 2017).

11   Deborah Bird Rose, 'Double Death', in *The Multispecies Salon: A Companion to the Book*, 2014, http://www.multispecies-salon.org/double-death/

12   Petra Buettner et al., 'Tick Paralysis in Spectacled Flying-Foxes (*Pteropus Conspicillatus*) in North Queensland, Australia: Impact of a Ground-Dwelling Ectoparasite Finding an Arboreal Host', *PLoS ONE* 8, no. 9 (2013).

13   Shane Knuth, accessed 2017, http://www.shaneknuth.com.au/hill/budget-for-bats/

14   Signage in the education centre at the Tolga Bat Hospital.

15   'Tolga scrub', Tolga Bat Hospital, https://tolgabathospital.org/about-us/tolga-scrub/

16   'Science and flying-foxes', Cairns Regional Council, http://www.cairns.qld.gov.au/community-environment/native-animals/flying-foxes/science-and-flying-foxes

17   Tolga Bat Hospital, https://tolgabathospital.org/about-us/

18   Interview with Jenny Mclean, 12 August 2010. All subsequent quotes are part of this extensive set of interviews, unless otherwise stated.

19   Interview with Mclean.

20   Signage from the Tolga Bat Hospital. To see video showing Chopper: see Natasha Fijn, 'Following Flying Foxes I: Can people really help flying foxes?', 2011, https://vimeo.com/31001848. This video was filmed by Natasha Fijn as part of this research project.

21   Surnames deleted for privacy.

22   Buettner et al., 'Tick Paralysis'.

23   Buettner et al., 'Tick Paralysis'.

24   Interview with Jenny Mclean, 12 August 2010; Buettner et al., 'Tick Paralysis'. The tick season in 2011 was especially severe, probably as a result of the effects of Cyclone Yasi. The Tolga hospital was unable to cope with the increased number of affected flying-foxes, and in November 2011 100 babies were airlifted south from Cairns to Brisbane where they were met by people who would foster them. In keeping with sound wildlife practice, the youngsters were sent back north to be released into the area from which they came.

25   Interview with Jenny Mclean, 12 August 2010. When I spoke to Jenny several years later, she reported that they no longer bury dead flying-foxes in this way. Instead, the local council prefers that they dispose of dead bats through their facility.

26   'Tick paralysis & cleft palate', Tolga Bat Hospital, https://tolgabathos pital.org/tick-paralysis-cleft-palate/

27   Booth et al., *Why NSW Should Ban*, 11.

28   Martin and McIlwee, 'Reproductive Biology', 98.

29   Martin and McIlwee, 'Reproductive Biology', 98.

30   Holland, '15,000 flying foxes killed in heat in Sydney'.

31   Hall and Richards, *Flying Foxes*, 49.

32   Interview with Louise Saunders, 12 August 2010.

33   David Fitzsimmons, 'Baby bats abandoned by starving mothers as food shortages kill the flying foxes', *Central Western Daily*, 23 November 2019, https://www.centralwesterndaily.com.au/story/4311890/baby-bat-rescues-reach-crisis-point/

34   Interview with Louise Saunders, 12 August 2010.

35   Cocos palm, *Syagrus romanzoffianum*, and the harm they cause to flying-foxes is detailed at: 'Cocos palms: A threat to flying-foxes', NSW

Office of Environment and Heritage, http://www.environment.nsw. gov.au/animals/cocos-palm-flying-foxes.htm

36   Pest Tales: Units, http://www.pestales.org.au/units.htm

37   This CRC completed its funding cycle and closed down in 2017. Much of the information is maintained on a new site: Centre for Invasive Species Solutions, https://invasives.com.au/. Pest Tales continues.

38   Definition at Pest Tales: Units, http://www.pestales.org.au/units.htm. When not offering innovative methods of killing, research focused on genetic engineering to eliminate reproductive capacity. Pest Tales was supported by the Invasive Animals CRC.

39   Donna Haraway, *Staying with the Trouble: Making Kin in the Chthulucene* (Durham, NC: Duke University Press, 2016), 14.

40   Denise Ford, 'An Historical Perspective of Changing Community Attitudes Towards Flying-foxes in Sydney', in *Managing the Grey-Headed Flying-Fox*, ed. Eby and Lunney, 146.

41   Ratcliffe, as quoted in Martin and McIlwee, 'Reproductive Biology', 104.

42   Martin and McIlwee, 'Reproductive Biology', 104.

43   Eby, *The Biology and Management*, 32.

44   Booth et al., *Why NSW Should Ban*, 5.

45   Toby Crockford, '"Wings shot off, babies clinging to dead mothers": Dozens of flying foxes killed on Sunshine Coast', *The Sydney Morning Herald*, 18 November 2017, https://www.smh.com.au/national/ queensland/wings-shot-off-babies-clinging-to-dead-mothers-dozens- of-flying-foxes-killed-on-sunshine-coast-20171118-p4yx2y.htm

46   Quoted from 'Teachers guide powerpoint presentation', Pest Tales, http://www.pestales.org.au/

47   Martin and McIlwee, 'Reproductive Biology', 105.

48   Eby and Lunney, *Managing the Grey-Headed Flying-Fox*.

49   Tim Pearson, 'Hard times for flying-foxes in Sydney', Step Inc: Community Based Environment Protection Since 1978, http://www. step.org.au/index.php/step-matters-issue-180/item/95-hard-times- for-flying-foxes-in-sydney

50   'Orchard flying fox netting program fully subscribed', NSW

Government: Department of Primary Industries, 17 March 2016, https://www.raa.nsw.gov.au/__data/assets/pdf_file/0009/598581/Orchard-Flying-Fox-netting-program-fully-subscribed.pdf

51 Hall and Richards, *Flying Foxes*, 105.

52 Peter Rigden, *To Net or Not to Net*, 3rd edn, Queensland Government: Department of Primary Industries and Fisheries, https://www.daf.qld.gov.au/__data/assets/pdf_file/0009/72954/Orchard-Netting-Report.pdf

53 According to Tim, this situation has changed over the last few years as a result of a combination of 'subsidised netting, a more difficult commercial situation, and many small peri-urban orchards being more profitably sold for housing development'.

54 Interview with Carol Booth, 10 August 2014.

55 David J. Paez et al., 'Optimal Foraging in Seasonal Environments: Implications for Residency of Australian Flying Foxes in Food-subsidized Urban Landscapes', *Philosophical Transactions of the Royal Society of London* 373 (2018).

56 Interview with Nick Edard, 25 June 2010.

57 'Booth v Bosworth ("Flying-fox case 1")', Environmental Defender's Office (QLD), https://www.edoqld.org.au/flying_fox_1

58 See, for example, Booth et al., *Why NSW Should Ban.*

59 Interview with Carol Booth, 10 August 2014.

60 This and the following information on pain, death and demographic impacts is sourced at Len Martin, 'Is the Fruit you Eat Flying-fox Friendly? The Effects of Orchard Electrocution Grids on Australian Flying-foxes (*Pteropus* spp., Megachiroptera)', in *The Biology and Management*, ed. Eby.

61 Martin and McIlwee, 'Reproductive Biology'.

62 Martin, 'Is the Fruit you Eat Flying-fox Friendly?', 387.

63 Martin, 'Is the Fruit you Eat Flying-fox Friendly?', 388.

64 Martin, 'Is the Fruit you Eat Flying-fox Friendly?', 388.

65 Martin, 'Is the Fruit you Eat Flying-fox Friendly?', 389.

66 Interview with Carol Booth, 10 August 2014.

67 'Booth v Bosworth'.

68 'Booth v Bosworth'.

69 'Booth v Yardley ("Flying-fox case 4")', Environmental Defender's Office (QLD), https://www.edoqld.org.au/flying_fox_4

70 Low, *Where Song Began*. Low is excellent on bird pollinators and on birds on the edge. Bees are insect pollinators which are suffering terribly under a range of impacts; see Freya Mathews, 'Planet Beehive', *Australian Humanities Review* 50 (2011).

71 Melissa Fyfe, 'The rainforest: Climate change icons under threat', *The Age*, 16 November 2005, http://rainforest-crc.jcu.edu.au/releases/the_rainforest.pdf

72 Calisher et al., 'Bats'.

73 Hume Field, 'The Role of Greyheaded Flying-Foxes in the Ecology of Hendra Virus, Menangle Virus and Australian Bat Lyssavirus', in *Managing the Grey-Headed Flying-Fox*, ed. Eby and Lunney, 139.

74 As quoted in Booth et al., *Why NSW Should Ban*, 17.

75 Field, 'The Role of Greyheaded Flying-Foxes'.

76 Field, 'The Role of Greyheaded Flying-Foxes'.

77 'Bat super immunity to lethal disease could help protect people', CSIRO, 23 February 2016, https://www.csiro.au/en/News/News-releases/2016/Bat-super-immunity-to-lethal-disease-could-help-protect-people

78 Raina K. Plowright et al., 'Urban Habituation, Ecological Connectivity and Epidemic Dampening: The Emergence of Hendra Virus from Flying Foxes (*Pteropus* spp.)', *Proceedings of the Royal Society B: Biological Sciences* 278, no. 1725 (2011).

79 Booth et al., *Why NSW Should Ban*, 17.

8 Cruelty and its allies

1 Francis Ratcliffe, *Flying Fox and Drifting Sand* 10.

2 Francis Ratcliffe, *The Flying Fox (Pteropus) in Australia* (Melbourne: Council for Scientific and Industrial Research, 1931), 46.

3 Geoffrey Bolton, *Spoils and Spoilers* (Sydney: George Allen & Unwin, 1981), 140.

4 Ratcliffe, *The Flying Fox (Pteropus) in Australia*.

5   Ratcliffe, *The Flying Fox (Pteropus) in Australia*, 4.

6   The Animal Studies Group, *Killing Animals* (Urbana: University of Illinois Press, 2006); Gary Steiner, *Animals and the Moral Community: Mental Life, Moral Status, and Kinship* (New York: Columbia University Press, 2008); Jeremy Bentham, *An Introduction to the Principles of Morals and Legislation* (1781).

7   Voiceless: The Animal Protection Institute, https://www.voiceless.org.au/

8   Elaine Scarry, *The Body in Pain: The Making and Unmaking of the World* (New York: Oxford University Press, 1985).

9   Daniel Lunney, Adele Reid, and Alison Matthews, 'Community Perceptions of Flying-foxes in New South Wales', in *Managing the Grey-Headed Flying-Fox*, ed. Eby and Lunney.

10   I find it hard to use the word 'host' to refer to a shock-jock, as the term implies a level of graciousness that is not characteristic of these people.

11   Pearson, 'Hard times for flying-foxes in Sydney'.

12   Julie Power, 'War on feral cats: Australia aims to cull 2 million', *The Sydney Morning Herald*, 17 February 2017, https://www.smh.com.au/national/war-on-feral-cats-australia-aims-to-cull-2-million-20170214-gucp4o.html

13   'Recovery planning: Restoring life to our threatened species', Australian Conservation Foundation, Birdlife Australia, Environmental Justice Australia, 2015.

14   The Invasive Animals website discusses some research-in-process on new methods of killing. With poisoning, the big problem is not the poison, per se, but delivering it to animals that are free-living in vast tracts of land.

15   Power, 'War on feral cats'.

16   J. M. Arthur, *The Default Country: A Lexical Cartography of Twentieth-Century Australia* (Sydney: University of New South Wales Press, 2003).

17   Power, 'War on feral cats'.

18   William Lynn, 'Australia's war on feral cats: Shaky science, missing ethics', *The Conversation*, 7 October 2015, https://theconversa

tion.com/australias-war-on-feral-cats-shaky-science-missing-ethics-47444

19   'Recovery planning: Restoring life to our threatened species', 2.

20   Thom van Dooren, 'Invasive Species in Penguin Worlds: An Ethical Taxonomy of Killing for Conservation', in *Conservation and Society* 9, no. 4 (2011).

21   R. I. Moore, *The Formation of a Persecuting Society: Authority and Deviance in Western Europe 950–1250*, 2nd edn (Malden, MA: Blackwell, 2007).

22   Moore, *Formation of a Persecuting Society*, 5 (emphasis added).

23   Moore, *Formation of a Persecuting Society*, 155.

24   See also Uli Linke, *Blood and Nation: European Aesthetics of Race* (Philadelphia: University of Pennsylvania Press, 1999).

25   Deborah Bird Rose, 'New World Poetics of Place: Along the Oregon Trail and in the National Museum of Australia', in *Rethinking 'Settler' Colonialism: History and Memory in Australia, Canada, Aotearoa New Zealand and South Africa*, ed. Annie Coombes (Manchester: Manchester University Press, 2006).

26   Alan Leishman, 'The history of Grey-headed flying-foxes in the Royal Botanic Gardens, Sydney' (2007), http://sydneybats.org.au/wpcon tent/uploads/2015/04/History_of_flying_foxes_Royal_Botanic_Gard ens_Sydney.pdf

27   G. C. Richards, 'The Development of Strategies for Management of the Flying-Fox Colony at the Royal Botanic Gardens, Sydney', in *Managing the Grey-Headed Flying-Fox*, ed. Eby and Lunney.

28   Richards, 'Development of Strategies', 189–99.

29   Cabramatta camp was the home of the first flying-foxes I encountered in Sydney (Chapter 1). It has since been abandoned.

30   Interview with Nick Edard, 25 June 2010. All following quotes are from this interview unless otherwise noted.

31   Tim Pearson, in an email dated Monday, 4 June 2012.

32   Tim Barlass, 'Botanic Gardens bats given their marching orders', *The Sydney Morning Herald*, 4 June 2012, https://www.smh.com.au/ environment/conservation/botanic-gardens-bats-given-their-march ing-orders-20120604-1zrwv.html

33 Fijn, 'Following Flying Foxes III'.

34 Lauren Berlant, *Cruel Optimism* (Durham, NC: Duke University Press, 2011), 243.

35 Adriana Cavarero, *Horrorism: Naming Contemporary Violence* (New York: Columbia University Press, 2011), 2.

36 For an excellent account, see Noel Castley-Wright, 'State of shame – Queensland's legislated animal cruelty', uploaded 6 April 2014, https://www.youtube.com/watch?v=0wF5D6k_9-U

37 See also van Dooren and Rose, 'Storied-Places in a Multispecies City'.

38 'Queensland councils will soon be allowed to chase off large numbers of flying-foxes from certain built-up areas without permission from the state. Currently local governments have to apply for a damage mitigation permit to disperse colonies of bats, which are protected in Australia. Councils say flying-fox colonies are distressing some communities which face long waits to secure dispersal permits, if they are granted at all. The Queensland government is changing the rules to allow councils in certain urban areas to scare away bat colonies without needing a permit, using non-lethal methods such as sound, light, smoke or chemicals.' See 'QLD councils given power to shift bats', News.com.au, 1 May 2013, https://www.news.com.au/national/breaking-news/qld-councils-given-power-to-shift-bats/news-story/d0987164c9289e04cd8a629e22e4f2ea

39 Interview with Louise Saunders, 12 August 2010.

40 Leslie Hall, 'Management of Flying-Fox Camps: What Have We Learnt in the Last Twenty five Years?' in *Managing the Grey-Headed Flying-Fox*, ed. Eby and Lunney.

41 Niza Yanay, *The Ideology of Hatred: The Psychic Power of Discourse* (New York: Fordham University Press, 2013), 6.

42 Achille Mbembe, 'Necropolitics', trans. Libby Meintjes, *Public Culture* 15, no. 1 (2003), 14; Michel Foucault, '17 March 1976', in *'Society must be defended': Lectures at the Collège de France, 1975–76*, ed. Mauro Bertani and Alessandro Fontana, trans. David Macey (London: Penguin, 2004), 239–63. For more on animals and biopolitics, see Dinesh Wadiwel, 'Biopolitics', in *Critical Terms for Animal Studies*, ed. Lori Gruen (Chicago: The University of Chicago Press, 2018).

43 David Lewis, 'Flying foxes should be destroyed: Katter', ABC,
    30 September 2011, https://www.abc.net.au/news/2011-09-30/
    flying-foxes-should-be-destroyed-katter/3193508

44 Peter Sloterdijk, 'Airquakes', *Environment and Planning D: Society
    and Space* 27, no. 1 (2009), 48–9.

45 Sloterdijk, 'Airquakes', 48.

46 Interview with Carol Booth, 10 August 2014.

47 As quoted in Zygmunt Bauman, *Modernity and the Holocaust* (Ithaca,
    NY: Cornell University Press, 1991), 208.

48 The CRC for Invasive Animals is the first port of call for learning
    about pests and death. From there, one goes to feral.org, and from
    there one enters the 'pest portal'. There one learns: 'The word "pest"
    is used to describe an animal that causes serious damage to a valued
    resource. Such a pest may be destructive, a nuisance, noisy or simply
    not wanted.' Pest Tales, http://www.pestales.org.au/units.htm

49 Deborah Bird Rose, 'Cosmopolitics: The Kiss of Life', *New Formations*
    76, no. 1 (2012); Deborah Bird Rose, 'Multispecies Knots of Ethical
    Time', *Environmental Philosophy* 9, no. 1 (2012).

50 Emmanuel Levinas, 'Useless Suffering', in *Entre-Nous: Thinking-of-
    the-Other*, trans. Michael Smith and Barbara Harshav (New York:
    Columbia University Press, 1998).

51 Sara Ahmed, 'The Non-Performativity of Anti-Racism', *Borderlands
    e-journal* 5, no. 3 (2005).

52 Ahmed, 'Non-Performativity'.

53 Zygmunt Bauman, 'The Holocaust's Life as a Ghost', in *The Holocaust's
    Life as a Ghost: Writing on Arts, Politics, Law and Education*, ed.
    F. C. Dacoste and Bernard Schwartz (Edmonton: University of Alberta
    Press, 2000).

9 Fidelity

1 The novelist Clarice Lispector is one of the people who tells us this:
   Clarice Lispector, *The Hour of the Star*, trans. Giovanni Pontiero (New
   York: New Directions Publishing, 1992), 1.

2 On onto-ecology see Cooke, 'What are the Animals Saying?' On

mutual life-giving see Plumwood, 'Shadow Places and the Politics of Dwelling'.

3   Freya Mathews, *For Love of Matter: A Contemporary Panpsychism* (Albany: State University of New York Press, 2003), 29; see also 171.

4   Plumwood, 'Nature in the Active Voice', 123.

5   Shestov equates G-d with life, earth and the joys of the ephemeral. I have left G-d out of my summary without, I hope, distorting the purpose of his analysis.

6   Shestov, *Athens and Jerusalem*, 180. This book was completed in 1937. Shestov died in 1938, at the age of seventy-two.

7   Levinas, 'Useless Suffering', 90.

8   Eileen Crist and Bruce H. Rinker, 'One Grand Organic Whole', in *Gaia in Turmoil*, ed. Crist and Rinker, 4–5.

9   Crist and Rinker, 'One Grand Organic Whole', 7.

10 Sagan, 'Life on a Margulisian Planet', 30.

11 Gregory Bateson, *Mind and Nature: A Necessary Unity* (Glasgow: Fontana/Collins, 1979).

12 Plumwood, *Feminism and the Mastery of Nature*.

13 Shestov, *Athens and Jerusalem*, 14.

14 Crist and Rinker, *Gaia in Turmoil*.

15 Isabelle Stengers, Erin Manning, and Brian Massumi, 'History through the Middle: Between Macro and Mesopolitics', trans. Brian Massumi, *Inflexions: A Journal for Research-Creation* 3 (2009), 6–7.

16 Stengers, Manning, and Massumi, 'History through the Middle', 6.

17 Isabelle Stengers, 'Beyond Conversation: The Risks of Peace', in *Process and Difference: Between Cosmological and Poststructuralist Postmodernisms*, ed. Catherine Keller and Anne Daniel (Albany: State University of New York Press, 2002), 241–5. See also, for example, Mark Lynas, *Six Degrees: Our Future on a Hotter Planet* (New York: HarperCollins, 2007).

18 The analysis here is parallel to J. K. Gibson-Graham's work with community economies and ways of breaking the hegemonic claims of capitalism. See J. K. Gibson-Graham, *The End of Capitalism (As We Knew It): A Feminist Critique of Political Economy* (Oxford: Blackwell Publishers, 1996).

NOTES | 263

19 Parts of this section are drawn from Rose, 'Cosmopolitics'.

20 K. A. Connell, U. Munro, and F. R. Torpy, 'Daytime Behaviour of the Grey-Headed Flying Fox *Pteropus Poliocephalus* Temminck (Pteropodidae: Megachiroptera) at an Autumn/Winter Roost', *Australian Mammalogy* 28, no. 1 (2006).

21 Connell, Munro, and Torpy, 'Daytime Behaviour'.

22 Justin Welbergen, 'The Grey-headed flying-fox, *Pteropus poliocephalus*', accessed 28 September 2010, http://www.zoo.cam.ac.uk/zoostaff/BBE/Welbergen/GHFlyingFox.htm

23 Summarised from Hall and Richards, *Flying Foxes*. Also see Martin and McIlwee, 'Reproductive Biology'.

24 Welbergen, 'The Grey-headed flying-fox'.

25 Leslie Hall, 'Management of Flying-Fox Camps', and Richards, 'The Development of Strategies for Management'.

26 Hall and Richards, *Flying Foxes*, 41–3.

27 Min Tan et al., 'Fellatio by Fruit Bats Prolongs Copulation Time', *PLoS ONE* 4, no. 10 (2009).

28 Tan et al., 'Fellatio by Fruit Bats'.

29 For example, Stephanie J. Smith and David M. Leslie, 'Pteropus Livingstonii', *Mammalian Species* 792 (2006), 4.

30 Hall and Richards, *Flying Foxes*, 45.

31 Smith and Leslie, 'Pteropus Livingstonii', 4.

32 Turner, 'When Species Kiss', 80.

33 Stengers, 'Beyond Conversation', 248 (emphasis added).

34 Edith Wyschogrod, 'Man-Made Mass Death: Shifting Concepts of Community', *Journal of the American Academy of Religion* 58, no. 2 (1990), 175.

35 Unpublished correspondence to members of Bat Conservation & Rescue QLD (BCRQ). Reproduced with permission of the author.

36 Unpublished account from article in the Bat Conservation and Rescue Newsletter. Quoted with permission of the author.

Bibliography

Abram, David. *The Spell of the Sensuous: Perception and Language in a More-Than-Human World*. New York: Vintage Books, 1996.

Ahmed, Sara. 'The Non-Performativity of Anti-Racism'. *Borderlands e-journal* 5, no. 3 (2005).

Apffel-Marglin, Frederique and Stephen Marglin. *Decolonizing Knowledge: From Development to Dialogue*. Oxford: Clarendon Press, 1996.

Arthur, J. M. *The Default Country: A Lexical Cartography of Twentieth-Century Australia*. Sydney: University of New South Wales Press, 2003.

Athanasiou, Tom. *Slow Reckoning*. Boston, MA: Little, Brown & Co., 1996.

Barad, Karen. *Meeting the Universe Halfway: Quantum Physics and the Entanglement of Matter and Meaning*. Durham, NC: Duke University Press, 2007.

Barad, Karen. 'Posthumanist Performativity: Toward an Understanding of How Matter Comes to Matter'. *Signs* 28, no. 3 (2003): 801–31.

Barlass, Tim. 'Botanic Gardens bats given their marching orders'. *The Sydney Morning Herald*, 4 June, 2012. https://www.smh.

com.au/environment/conservation/botanic-gardens-bats-given-their-marching-orders-20120604-1zrwv.html

Bateson, Gregory. *Mind and Nature: A Necessary Unity*. Glasgow: Fontana/Collins, 1979.

Bauman, Zygmunt. *Modernity and the Holocaust*. Ithaca, NY: Cornell University Press, 1991.

Bauman, Zygmunt. 'The Holocaust's Life as a Ghost'. In *The Holocaust's Life as a Ghost: Writings on Art, Politics, Law and Education*, edited by F. C. Decoste and Bernard Schwartz, 3–15. Edmonton: University of Alberta, 2000.

Bavas, Josh. 'About 100,000 bats dead after heatwave in southern Queensland'. ABC, 8 January 2014. http://www.abc.net.au/news/2014-01-08/hundred-thousand-dead-bats-after-qld-heat wave-rspca-says/5190644

Bekoff, Marc. *The Animal Manifesto: Six Reasons for Expanding Our Compassion Footprint*. Novato, CA: New World Library, 2010.

Bentham, Jeremy. *An Introduction to the Principles of Morals and Legislation*. 1781.

Berlant, Lauren. *Cruel Optimism*. Durham, NC: Duke University Press, 2011.

Bolton, Geoffrey. *Spoils and Spoilers*. Sydney: George Allen & Unwin, 1981.

Bond, W. J. 'Assessing the Risk of Plant Extinction Due to Pollinator and Disperser Failure'. In *Extinction Rates*, edited by John Lawton and Robert May, 131–46. Oxford: Oxford University Press, 1995.

Booth, Carol et al. *Why NSW Should Ban the Shooting of Flying Foxes*. Sydney: Humane Society International, 2008.

Bowman, D. M. J. S. 'Tansley Review No. 101: The Impact of Aboriginal Landscape Burning on the Australian Biota'. *New Phytologist* 140 (1998): 385–410.

Buchanan, Brett. *Onto-Ethologies: The Animal Environments of Uexküll, Heidegger, Merleau-Ponty, and Deleuze*. Albany: State University of New York Press, 2008.

Buettner, Petra et al. 'Tick Paralysis in Spectacled Flying-Foxes (*Pteropus Conspicillatus*) in North Queensland, Australia: Impact of a Ground-Dwelling Ectoparasite Finding an Arboreal Host'. *PLoS ONE* 8, no. 9 (2013): 1–10.

Burd, Martin. 'Colourful language – it's how Aussie birds and flowers "speak"'. *The Conversation*, 26 February 2014. https://theconversation.com/colourful-language-its-how-aussie-birds-and-flowers-speak-23659

Butler, Judith. *Precarious Life: The Powers of Mourning and Violence*. London: Verso, 2004.

Calisher, Charles et al. 'Bats: Important Reservoir Hosts of Emerging Viruses'. *Clinical Microbiology Reviews* 19, no. 3 (2006): 531–45.

Carroll, Sean. *Endless Forms Most Beautiful: The New Science of Evo Devo and the Making of the Animal Kingdom*. London: Quercus, 2011.

Castley-Wright, Noel. 'State of shame – Queensland's legislated animal cruelty'. Uploaded 6 April 2014. https://www.youtube.com/watch?v=0wF5D6k_9-U

Cavarero, Adriana. *Horrorism: Naming Contemporary Violence*. New York: Columbia University Press, 2011.

Clark, David. 'On Being "the Last Kantian in Nazi Germany": Dwelling with Animals after Levinas'. In *Animal Acts: Configuring the Human in Western History*, edited by Jennifer Ham and Matthew Senior, 165–98. New York: Routledge, 1999.

Coetzee, J. M. *The Lives of Animals*. Princeton, NJ: Princeton University Press, 2001.

Connell, K. A., U. Munro, and F. R. Torpy. 'Daytime Behaviour of the Grey-Headed Flying Fox *Pteropus Poliocephalus* Temminck (Pteropodidae: Megachiroptera) at an Autumn/Winter Roost'. *Australian Mammalogy* 28, no. 1 (2006): 7–14.

Cooke, Stuart. 'What are the Animals Saying?' *Plumwood Mountain: An Australian Journal of Ecopoetry and Ecopoetics*. https://plumwoodmountain.com/what-are-the-animals-saying/

Crist, Eileen. *Images of Animals: Anthropomorphism and Animal Mind*. Philadelphia, PA: Temple University Press, 1999.

Crist, Eileen. 'Intimations of Gaia'. In *Gaia in Turmoil*, edited by Crist and Rinker, 315–34.

Crist, Eileen and Bruce H. Rinker. 'One Grand Organic Whole'. In *Gaia in Turmoil*, edited by Crist and Rinker, 3–20.

Crist, Eileen and Bruce H. Rinker, eds. *Gaia in Turmoil: Climate Change, Biodepletion, and Earth Ethics in an Age of Crisis.* Cambridge, MA: MIT Press, 2010.

Crockford, Toby. '"Wings shot off, babies clinging to dead mothers": Dozens of flying foxes killed on Sunshine Coast'. *The Sydney Morning Herald*, 18 November 2017. https://www.smh.com. au/national/queensland/wings-shot-off-babies-clinging-to-dead-mothers-dozens-of-flying-foxes-killed-on-sunshine-coast-2017 1118-p4yx2y.htm

Danaiyarri, Hobbles. 'The Saga of Captain Cook'. In *Australia's Empire*, edited by Deryck Schreuder and Stuart Ward, 27–32. Oxford: Oxford University Press, 2008.

de Waal, Frans. *Are We Smart Enough to Know How Smart Animals Are?* New York: W. W. Norton, 2016.

de Waal, Frans. 'Putting the Altruism Back into Altruism: The Evolution of Empathy'. *Annual Review of Psychology* 59 (2008): 279–300.

Derrida, Jacques. *The Animal That Therefore I Am*. Edited by Marie-Louise Mallet. Translated by David Wills. New York: Fordham University Press, 2008.

Didion, Joan. *The White Album*. New York: Farrar, Strauss & Giroux, 1990.

Dietrich, William. *The Final Forest: The Battle for the Last Great Trees of the Pacific Northwest*. New York: Penguin Books, 1992.

Eby, Peggy. 'Seasonal Movements of Grey-Headed Flying-Foxes, *Pteropus Poliocephalus* (Chiroptera: Pteropodidae) from Two Maternity Camps in Northern New South Wales'. *Wildlife Research* 18 (1991): 547–59.

Eby, Peggy, ed. *The Biology and Management of Flying Foxes in NSW*. Sydney: NSW National Parks and Wildlife Service, 1995.

Eby, Peggy and Daniel Lunney, eds. *Managing the Grey-Headed*

Flying-Fox as a Threatened Species in New South Wales. Sydney: Royal Zoological Society of New South Wales, 2002.

Edelglass, William, James Hatley, and Christian Diehm, eds. *Facing Nature: Levinas and Environmental Thought*. Pittsburgh, PA: Duquesne University Press, 2012.

Elkin, A. P. *The Australian Aborigines: How to Understand Them*. Sydney: Angus and Robertson, 1954.

Ellis, Catherine. *Aboriginal Music, Education for Living: Cross Cultural Experiences from South Australia*. St Lucia: University of Queensland Press, 1985.

Field, Hume. 'The Role of Greyheaded Flying-Foxes in the Ecology of Hendra Virus, Menangle Virus and Australian Bat Lyssavirus'. In *Managing the Grey-Headed Flying-Fox*, edited by Eby and Lunney, 139–41.

Fijn, Natasha. 'Following Flying Foxes I: Can people really help flying foxes?' 2011. https://vimeo.com/31001848

Fijn, Natasha. 'Following Flying Foxes II: What is it like to foster an orphan?' 2011. https://vimeo.com/30877202

Fijn, Natasha. 'Following Flying Foxes III: What happens when humans decide on zero tolerance?' 2011. https://vimeo.com/30881933

Fitzsimmons, David. 'Baby bats abandoned by starving mothers as food shortages kill the flying foxes'. *Central Western Daily*, 23 November 2019. https://www.centralwesterndaily.com.au/story/4311890/baby-bat-rescues-reach-crisis-point/

Flanagan, Richard. 'Opening Address for "Janet Laurence: After Eden"'. 2012.

Ford, Denise. 'An Historical Perspective of Changing Community Attitudes Towards Flying-foxes in Sydney'. In *Managing the Grey-Headed Flying-Fox*, edited by Eby and Lunney, 146–59.

Foucault, Michel. '17 March 1976'. In *'Society must be defended': Lectures at the Collège de France, 1975–76*, 239–63. Edited by Mauro Bertani and Alessandro Fontana. Translated by David Macey. London: Penguin, 2004.

Fyfe, Melissa. 'The rainforest: Climate change icons under threat'.

The Age, 16 November 2005. http://rainforestcrc.jcu.edu.au/releases/the_rainforest.pdf

Gagliano, Monica. *Thus Spoke the Plant: A Remarkable Journey of Groundbreaking Scientific Discoveries and Personal Encounters with Plants*. Berkeley, CA: North Atlantic Books, 2018.

Gammage, Bill. *The Biggest Estate on Earth*. Sydney: Allen & Unwin, 2012.

Geertz, Clifford. 'Ethos, World View, and the Analysis of Sacred Symbols'. In *The Interpretation of Cultures*, 126–41. New York: Basic Books, 1996.

Giannini, Norberto P. and Nancy B. Simmons. 'A Phylogeny of Megachiropteran Bats (Mammalia: Chiroptera: Pteropodidae) Based on Direct Optimization Analysis of One Nuclear and Four Mitochondrial Genes'. *Cladistics* 19, no. 6 (2003): 496–511.

Gibson-Graham, J. K. *The End of Capitalism (As We Knew It): A Feminist Critique of Political Economy*. Oxford: Blackwell Publishers, 1996.

Graham, Mary. 'Some Thoughts on the Philosophical Underpinnings of Aboriginal Worldviews'. *Australian Humanities Review* 45 (2008): 181–94.

Grinde, Donald and Bruce Johansen. *Ecocide of Native America: Environmental Destruction of Indian Lands and Peoples*. Santa Fe, NM: Clear Light Publishers, 1995.

Hall, Leslie. 'Management of Flying-Fox Camps: What Have we Learnt in the Last Twenty-five Years?' In *Managing the Grey-Headed Flying-Fox*, edited by Eby and Lunney, 215–24.

Hall, Leslie and Greg Richards. *Flying Foxes: Fruit and Blossom Bats of Australia*. Sydney: University of New South Wales Press, 2000.

Hall, Matthew. *Plants as Persons: A Philosophical Botany*. Albany: State University of New York Press, 2011.

Hannam, Peter. 'Records melt in our hottest year'. *The Sydney Morning Herald*, 21 December 2013. https://www.smh.com.au/environment/climate-change/records-melt-in-our-hottest-year-20131220-2zqrt.html

Haraway, Donna. 'Situated Knowledges: The Science Question

in Feminism and the Privilege of Partial Perspective'. *Feminist Studies* 14, no. 3 (1988): 575–99.

Haraway, Donna. *Staying with the Trouble: Making Kin in the Chthulucene.* Durham, NC: Duke University Press, 2016.

Haraway, Donna. *When Species Meet.* Minneapolis: University of Minnesota Press, 2008.

Harding, Stephan. 'Gaia and Biodiversity'. In *Gaia in Turmoil*, edited by Crist and Rinker, 107–24.

Harding, Stephan and Lynn Margulis. 'Water Gaia: 3.5 Thousand Million Years of Wetness on Planet Earth'. In *Gaia in Turmoil*, edited by Crist and Rinker, 41–59.

Harries-Jones, Peter. *A Recursive Vision: Ecological Understanding and Gregory Bateson.* Toronto: University of Toronto Press, 1995.

Harvey, Graham. *Animism: Respecting the Living World.* New York: Columbia University Press, 2006.

Harvey, Graham, ed. *The Handbook of Contemporary Animism.* Durham: Acumen, 2013.

Hatley, James. *Suffering Witness: The Quandary of Responsibility after the Irreparable.* Albany: State University of New York Press, 2000.

Hatley, James. 'The Anarchical Goodness of Creation: Monotheism in Another's Voice'. In *Facing Nature*, edited by Edelglass, Hatley, and Diehm, 253–78.

Hatley, James. 'The Virtue of Temporal Discernment: Rethinking the Extent and Coherence of the Good in a Time of Mass Species Extinction'. *Environmental Philosophy* 9, no. 1 (2012): 1–22.

Herzfeld, Michael. *Anthropology: Theoretical Practice in Culture and Society.* Malden and Oxford: Blackwell Publishers, 2001, 283–4.

Hoffmeyer, Jesper, ed. *A Legacy for Living Systems: Gregory Bateson as Precursor to Biosemiotics.* New York: Springer, 2008.

Hoffmeyer, Jesper. 'From Thing to Relation. On Bateson's Bioanthropology'. In *A Legacy for Living Systems*, edited by Hoffmeyer, 27–44.

Hoffmeyer, Jesper. *Signs of Meaning in the Universe*. Translated by Barbara Haveland. Bloomington: Indiana University Press, 1993.

Holland, Malcolm. '15,000 flying foxes killed in heat in Sydney'. *The Daily Telegraph*, 23 January 2013. https://www.dailytele graph.com.au/flying-foxes-killed-in-heat-in-sydney/news-story/ 86dc7bfeb25ee5cb0611e6534e5c9d62

Holland, Maximilian. *Social Bonding and Nurture Kinship: Compatibility between Cultural and Biological Approaches*. North Charleston, SC: CreateSpace Press, 2012.

Huth, John Edward. *The Lost Art of Finding Our Way*. Cambridge, MA: The Belknap Press of Harvard University Press, 2013.

Janzen, Dean. 'The Deflowering of Central America'. *Natural History* 83, no. 4 (1974): 49–53.

Jonas, Hans. 'The Burden and Blessing of Mortality'. *Hastings Center Report* 22, no. 1 (1992): 34–40.

Kirksey, Eben and Stefan Helmreich. 'The Emergence of Multispecies Ethnography'. *Cultural Anthropology* 25, no. 4 (2010): 545–76.

Laird, Tessa. *Bat*. London: Reaktion Books, 2018.

Larsen, Elizabeth, Marjorie Beck, Elizabeth Hartnell, and Michael Creenaune. 'Neighbours of Ku-ring-gai Flying-fox Reserve: Community Attitudes Survey 2001'. In *Managing the Grey-Headed Flying-Fox*, edited by Eby and Lunney, 225–39.

Leishman, Alan. 'The history of Grey-headed flying-foxes in the Royal Botanic Gardens, Sydney'. 2007. http://www.sydneybats. org.au/cms/index.php?id=11,61,0,0,1,0

Lestel, Dominique. 'Eprouver La Personne Comme Personnage'. In *Personne/Personnage*, edited by Thierry Lenain and Aline Wiame, 123–37. Paris: Vrin, 2011.

Lestel, Dominique. *L'Animal est l'avenir de l'homme: Munitions pour ceux qui veulent (toujours) défendre les animaux*. Paris: Fayard, 2010.

Lestel, Dominique. 'La Haine de l'animal'. In *Aux origines de l'environnement*, edited by Pierre-Henri Gouyon and Hélène Leriche, 192–205. Paris: Fayard, 2010.

Lestel, Dominique. 'The Question of the Animal Subject: Thoughts on the Fourth Wound to Human Narcissism'. Translated by Hollis Taylor. *Angelaki* 19, no. 3 (2014): 113–25.

Lestel, Dominique, Florence Brunois, and Florence Gaunet. 'Etho-Ethnology and Ethno-Ethology'. *Social Science Information* 45, no. 2 (2006): 155–77.

Levinas, Emmanuel. *Difficult Freedom: Essays on Judaism.* Translated by Sean Hand. Baltimore, MD: Johns Hopkins University Press, 1997.

Levinas, Emmanuel. 'The Paradox of Morality: An Interview with Emmanuel Levinas'. In *The Provocation of Levinas: Rethinking the Other*, edited by Robert Bernasconi and David Woods, 168–80. London: Routledge, 1988.

Levinas, Emmanuel. *Totality and Infinity: An Essay on Exteriority.* Translated by Alphonso Lingis. Pittsburgh, PA: Duquesne University Press, 1969.

Levinas, Emmanuel. 'Useless Suffering'. In *Entre-Nous: Thinking-of-the-Other*, 91–101. Translated by Michael Smith and Barbara Harshav. New York: Columbia University Press, 1998.

Levinas, Emmanuel and Richard Kearney. 'Dialogue with Emmanuel Levinas'. In *Face to Face with Levinas*, edited by Richard Cohen, 13–33. Albany: State University of New York Press, 1986.

Lévy-Bruhl, Lucien. *How Natives Think.* Translated by Lilian Clare. London: George Allen & Unwin Ltd, 1926.

Lévy-Bruhl, Lucien. *Primitive Mythology: The Mythic World of the Australian and Papuan Natives.* Translated by Brian Elliott. St Lucia: Queensland University Press, 1983.

Lewis, Darrell. *Slower Than the Eye Can See.* Darwin: Tropical Savannas CRC, 2002.

Lewis, Darrell and Deborah Bird Rose. *The Shape of the Dreaming: The Cultural Significance of Victoria River Rock Art.* Canberra: Aboriginal Studies Press, 1988.

Lewis, David. 'Flying foxes should be destroyed: Katter'. ABC, 30 September 2011. https://www.abc.net.au/news/2011-09-30/flying-foxes-should-be-destroyed-katter/3193508

Linke, Uli. *Blood and Nation: European Aesthetics of Race.* Philadelphia: University of Pennsylvania Press, 1999.

Lispector, Clarice. *The Hour of the Star.* Translated by Giovanni Pontiero. New York: New Directions Publishing, 1992.

Low, Tim, *The New Nature.* Melbourne: Viking, 2002.

Low, Tim. *Where Song Began: Australia's Birds and How They Changed the World.* Sydney: Penguin Random House Australia, 2014.

Lunney, Daniel, Adele Reid, and Alison Matthews. 'Community Perceptions of Flying-foxes in New South Wales'. In *Managing the Grey-Headed Flying-Fox,* edited by Eby and Lunney, 160–75.

Lynas, Mark. *Six Degrees: Our Future on a Hotter Planet.* New York: HarperCollins, 2007.

Lynn, William. 'Australia's war on feral cats: Shaky science, missing ethics'. *The Conversation,* 7 October 2015. https://theconversa tion.com/australias-war-on-feral-cats-shaky-science-missing-ethics-47444

Macdonald, David and M. Karen Laurenson. 'Infectious Disease: Inextricable Linkages between Human and Ecosystem Health'. *Biological Conservation* 131 (2006): 143–50.

McKnight, David. 'Men, Women, and Other Animals: Taboo and Purification among the Wikmungkan'. In *The Interpretation of Symbolism,* edited by Roy Willis, 77–97. London: Malaby Press, 1975.

Magowan, Fiona. *Melodies of Mourning: Music and Emotion in Northern Australia.* Nedlands: University of Western Australia Press, 2007.

Margulis, Lynn. *Symbiotic Planet: A New Look at Evolution.* Amherst, MA: Basic Books, 1998.

Margulis, Lynn and Dorion Sagan. *What Is Life?* Berkeley: University of California Press, 2000.

Markus, Nicola and Les Hall. 'Foraging Behaviour of the Black Flying-Fox (*Pteropus Alecto*) in the Urban Landscape of Brisbane, Queensland'. *Wildlife Research* 31, no. 3 (2004): 345–55.

Martin, Len. 'Is the Fruit you Eat Flying-fox Friendly? The Effects of Orchard Electrocution Grids on Australian Flying-foxes (*Pteropus* spp., Megachiroptera)'. In *The Biology and Management of Flying Foxes in NSW*, edited by Eby, 380–90.

Martin, L. and A. P. McIlwee. 'The Reproductive Biology and Intrinsic Capacity for Increase of the Grey-Headed Flying-Fox Poliocephalus (Megachiroptera), and the Implications of Culling'. In *Managing the Grey-Headed Flying-Fox*, edited by Eby and Lunney, 91–108.

Mathews, Freya. 'Ceres: Singing up the City'. *PAN: Philosophy, Activism, Nature* no. 1 (2000): 5–15.

Mathews, Freya. 'Earth as Ethic'. In *Manifesto for Living in the Anthropocene*, edited by Katherine Gibson, Deborah Bird Rose, and Ruth Fincher, 91–5. Brooklyn, NY: Punctum, 2015.

Mathews, Freya. 'From Biodiversity-based Conservation to an Ethic of Bio-proportionality'. *Biological Conservation* 200 (2016): 140–8.

Mathews, Freya. *For Love of Matter: A Contemporary Panpsychism.* Albany: State University of New York Press, 2003.

Mathews, Freya. 'Planet Beehive'. *Australian Humanities Review* 50 (2011): 159–78.

Mathews, Freya. *The Ecological Self.* London: Routledge, 1991.

Mbembe, Achille. 'Necropolitics'. Translated by Libby Meintjes. *Public Culture* 15, no. 1 (2003): 11–40.

Mohr, Tony. 'Weather on steroids: Climate change in action'. ABC, 7 January 2013. http://www.abc.net.au/news/2013-01-07/mohr-weather-on-steroids/4455022

Moore, R. I. *The Formation of a Persecuting Society: Authority and Deviance in Western Europe 950–1250*, 2nd edn. Malden, MA: Blackwell, 2007.

Morphy, Howard. 'From Dull to Brilliant: The Aesthetics of Spiritual Power among the Yolngu'. *Man*, New Series, 24, no. 1 (1989): 21–40.

Nagel, Thomas. 'What Is It Like to Be a Bat?' *The Philosophical Review* 83, no. 4 (1974): 435–50.

Newton, Adam Zachary. *Narrative Ethics*. Cambridge, MA: Harvard University Press, 1995.

Paez, David J., Olivier Restif, Peggy Eby, and Raina K. Plowright. 'Optimal Foraging in Seasonal Environments: Implications for Residency of Australian Flying Foxes in Food-subsidized Urban Landscapes'. *Philosophical Transactions of the Royal Society of London* 373 (2018): 1–8.

Parris, Kristen and D. L. Hazell. 'Biotic Effects of Climate Change in Urban Environments: The Case of the Grey-Headed Flying-Fox (*Pteropus poliocephalus*) in Melbourne, Australia'. *Biological Conservation* 124 (2005): 267–76.

Parry-Jones, K. A. and M. L. Augee. 'Food Selection by Grey-Headed Flying Foxes (*Pteropus Poliocephalus*) Occupying a Summer Colony Site near Gosford, New South Wales'. *Wildlife Research* 18, no. 1 (1991): 111–24.

Parry-Jones, K. A. and M. L. Augee. 'Movements of Grey-Headed Flying Foxes (*Pteropus Poliocephalus*) to and from a Colony Site on the Central Coast of New South Wales'. *Wildlife Research* 19 (1992): 331–40.

Parsons, Jennifer, Simon Robson, and Louise Shilton. 'Roost Fidelity in Spectacled Flying-Foxes *Pteropus conspicullatus*: Implications for Conservation and Management'. In *The Biology and Conservation of Australasian Bats*, edited by Bradley Law, Peggy Eby, Daniel Lunney, and Lindy Lumsden, 66–71. Sydney: Royal Zoological Society of New South Wales, 2011.

Pascoe, Bruce. *Dark Emu*. Broome: Magabala Books, 2014.

Pearson, Tim. 'Hard times for flying-foxes in Sydney'. Step Inc: Community Based Environment Protection Since 1978. http://www.step.org.au/index.php/step-matters-issue-180/item/95-hard-times-for-flying-foxes-in-sydney

Perry, Dan. '"Endemic Aliens": Grey-Headed Flying-Foxes at the Melbourne Royal Botanic Gardens'. *Journal of Animal Ethics* 2, no. 2 (2012): 162–78.

Pettigrew, John. 'Are Flying Foxes Really Primates?' *Bats Magazine* 3, no. 2 (Summer 1986).

Plowright, Raina K. et al. 'Urban Habituation, Ecological Connectivity and Epidemic Dampening: The Emergence of Hendra Virus from Flying Foxes (*Pteropus* spp.)'. *Proceedings of the Royal Society of Biological Sciences* 278, no. 1725 (2011): 3703–12.

Plumwood, Val. *Environmental Culture: The Ecological Crisis of Reason.* London: Routledge, 2002.

Plumwood, Val. *Feminism and the Mastery of Nature.* London: Routledge, 1993.

Plumwood, Val. 'Nature in the Active Voice'. *Australian Humanities Review* 46 (2009): 113–29.

Plumwood, Val. 'Shadow Places and the Politics of Dwelling'. *Australian Humanities Review* 44 (2008): 139–50.

Povinelli, Elizabeth. *Labor's Lot: The Power, History, and Culture of Aboriginal Action.* Chicago: University of Chicago Press, 1993.

Power, Julie. 'War on feral cats: Australia aims to cull 2 million'. *The Sydney Morning Herald*, 17 February 2017. https://www.smh.com.au/national/war-on-feral-cats-australia-aims-to-cull-2-million-20170214-gucp4o.html

Prigogine, Ilya. *The End of Certainty: Time, Chaos and the New Laws of Nature.* New York: The Free Press, 1997.

Puig de la Bellacasa, Maria. '"Nothing comes without its world": Thinking with Care'. *The Sociological Review* 60, no. 2 (2012): 197–216.

Radcliffe-Brown, A. R. 'The Rainbow-Serpent Myth in South-East Australia'. *Oceania* 1, no. 3 (1930): 342–47.

Ratcliffe, Francis. *Flying Fox and Drifting Sand: The Adventures of a Biologist in Australia.* Sydney: Angus and Robertson, 1948.

Ratcliffe, Francis. *The Flying Fox (Pteropus) in Australia.* Melbourne: Council for Scientific and Industrial Research, 1931.

Reznikova, Zhanna. *Animal Intelligence: From Individual to Social Cognition.* Cambridge: Cambridge University Press, 2007.

Richards, G. C. 'The Development of Strategies for Management of the Flying-Fox Colony at the Royal Botanic Gardens, Sydney'.

In *Managing the Grey-Headed Flying-Fox*, edited by Eby and Lunney, 196–201.

Rigden, Peter. *To Net or Not to Net*, 3rd edn. Queensland Government: Department of Primary Industries and Fisheries, 2008. https://www.daf.qld.gov.au/__data/assets/pdf_file/0009/7 2954/Orchard-Netting-Report.pdf

Roberts, Billie J., Carla P. Catterall, Peggy Eby, and John Kanowski. 'Long-Distance and Frequent Movements of the Flying-Fox *Pteropus poliocephalus*: Implications for Management'. *PLoS ONE* 7, no. 8 (2012): e42532.

Rose, Deborah Bird. 'Conflict Resolution and Decolonisation: Aboriginal Australian Case Studies in "Enlarged Thinking"'. In *Mediating across Difference: Oceanic and Asian Approaches to Conflict Resolution*, edited by Morgan Brigg and Roland Bleiker, 100–14. Honolulu: University of Hawaii Press, 2011.

Rose, Deborah Bird. 'Cosmopolitics: The Kiss of Life'. *New Formations* 76, no. 1 (2012): 101–13.

Rose, Deborah Bird. 'Country for Yarralin'. 7 July 2016. https://webarchive.nla.gov.au/tep/177305

Rose, Deborah Bird. 'Death and Grief in a World of Kin'. In *The Handbook of Contemporary Animism*, edited by Harvey, 137–47.

Rose, Deborah Bird. *Dingo Makes Us Human: Life and Land in an Australian Aboriginal Culture*. Cambridge: Cambridge University Press, 1992.

Rose, Deborah Bird. 'Double Death'. In *The Multispecies Salon: A Companion to the Book*. 2014. http://www.multispecies-salon.org/double-death/

Rose, Deborah Bird. 'Flying-foxes at Port Keats 2013'. https://vimeo.com/75266158

Rose, Deborah Bird. 'Gendered Substance and Objects in Ritual: An Australian Study'. *Material Religion: The Journal of Objects, Art and Belief* 3, no. 1 (2007): 3–47.

Rose, Deborah Bird. *Hidden Histories: Black Stories from Victoria River Downs, Humbert River, and Wave Hill Stations*. Canberra: Aboriginal Studies Press, 1991.

Rose, Deborah Bird. 'Histories and Rituals: Land Claims in the Territory'. In *The Age of Mabo: History, Aborigines and Australia*, edited by Bain Attwood, 35–53. Sydney: Allen & Unwin, 1996.

Rose, Deborah Bird. 'Multispecies Knots of Ethical Time'. *Environmental Philosophy* 9, no. 1 (2012): 127–40.

Rose, Deborah Bird. 'New World Poetics of Place: Along the Oregon Trail and in the National Museum of Australia'. In *Rethinking 'Settler' Colonialism: History and Memory in Australia, Canada, Aotearoa New Zealand and South Africa*, edited by Annie Coombes, 228–44. Manchester: Manchester University Press, 2006.

Rose, Deborah Bird, *Nourishing Terrains: Australian Aboriginal Views of Landscape and Wilderness*. Canberra: Australian Heritage Commission, 1996.

Rose, Deborah Bird. 'On History, Trees and Ethical Proximity'. *Postcolonial Studies* 11, no. 2 (2008): 157–67.

Rose, Deborah Bird. 'Remembrance, in the wake of suicide'. 28 March 2016. https://webarchive.nla.gov.au/tep/177305

Rose, Deborah Bird. *Reports from a Wild Country: Ethics for Decolonisation*. Sydney: University of New South Wales Press, 2004.

Rose, Deborah Bird. 'The Saga of Captain Cook: Morality in Aboriginal and European Law'. *Australian Aboriginal Studies* 2 (1984): 24–39.

Rose, Deborah Bird. 'Val Plumwood's Philosophical Animism'. *Environmental Humanities* 3, no. 1 (2013): 93–109.

Rose, Deborah Bird. *Wild Dog Dreaming: Love and Extinction*. Charlottesville: University of Virginia Press, 2011.

Rose, Deborah Bird, Thom van Dooren, and Matthew Chrulew. 'Introduction: Telling Extinction Stories'. In *Extinction Studies: Stories of Death, Time and Generations*, edited by Deborah Bird Rose, Thom van Dooren, and Matthew Chrulew, 1–17. New York: Columbia University Press, 2017.

Rose, Deborah Bird, with Sharon D'Amico, Nancy Daiyi, Kathy

Deveraux, Margy Daiyi, Linda Ford, and April Bright. *Country of the Heart: An Indigenous Australian Homeland*. Canberra: Aboriginal Studies Press, 2002.

Roth, John. *Ethics after the Holocaust: Perspectives, Critiques, and Responses*. St. Paul, MN: Paragon House, 1999.

Rowlands, Mark. *Can Animals Be Moral?* Oxford: Oxford University Press, 2012.

Sagan, Dorion. 'Life on a Margulisian Planet: A Son's Philosophical Reflections'. In *Earth, Life, and System: Evolution and Ecology on a Gaian Planet*, edited by Bruce Clarke, 13–38. New York: Fordham University Press, 2015.

Scarry, Elaine. *The Body in Pain: The Making and Unmaking of the World*. New York: Oxford University Press, 1985.

Schneider, David. *A Critique of the Study of Kinship*. Ann Arbor: University of Michigan Press, 1984.

Scott, James C. *Seeing Like a State: How Certain Schemes to Improve the Human Condition Have Failed*. New Haven, CT: Yale University Press, 1998.

Scott, James C. *The Art of Not Being Governed: An Anarchist History of Upland Southeast Asia*. New Haven, CT: Yale University Press, 2010.

Shaw, Scott Richard. *Planet of the Bugs*. Chicago: University of Chicago Press, 2014.

Shepard, Paul. *The Others: How Animals Made Us Human*. Washington, DC: Island Press, 1996.

Shestov, Lev. *Athens and Jerusalem*. Translated by Bernard Martin. New York: Simon and Schuster, 1968.

Shestov, Lev. 'Children and Stepchildren of Time: Spinoza in History'. In *A Shestov Anthology*, edited by Bernard Martin, 215–43. Athens: Ohio University Press, 1970.

Shestov, Lev. 'Myth and Truth: On the Metaphysics of Knowledge'. In *Speculation and Revelation*, 118–29. Translated by Bernard Martin. Athens: Ohio University Press, 1982.

Shestov, Lev. 'Speculation and Apocalypse: The Religious Philosophy of Vladimir Solovyov'. In *Speculation and Revelation*,

18–88. Translated by Bernard Martin. Athens: Ohio University Press, 1982.

Shestov, Lev. *Speculation and Revelation*. Translated by Bernard Martin. Athens: Ohio University Press, 1982.

Shilton, L. A. et al. 'Landscape-Scale Redistribution of a Highly Mobile Threatened Species, *Pteropus Conspicillatus* (Chiroptera, Pteropodidae), in Response to Tropical Cyclone Larry'. *Austral Ecology* 33, no. 4 (2008): 549–61.

Simmons, Nancy B. 'Taking Wing'. *Scientific American* 229, no. 6 (2008): 96–103.

Sloterdijk, Peter. 'Airquakes'. *Environment and Planning D: Society and Space* 27, no. 1 (2009): 41–57.

Smith, Stephanie J. and David M. Leslie. 'Pteropus Livingstonii'. *Mammalian Species* 792 (2006): 1–5.

Spencer, H. J., C. Palmer, and K. Parry-Jones. 'Movements of Fruit-Bats in Eastern Australia, Determined by Using Radio-Tracking'. *Wildlife Research* 18, no. 4 (1991): 463–67.

Steiner, Gary. *Animals and the Moral Community: Mental Life, Moral Status, and Kinship*. New York: Columbia University Press, 2008.

Stengers, Isabelle. 'Beyond Conversation: The Risks of Peace'. In *Process and Difference: Between Cosmological and Poststructuralist Postmodernisms*, edited by Catherine Keller and Anne Daniel, 235–55. Albany: State University of New York Press, 2002.

Stengers, Isabelle. *Cosmopolitics I*. Translated by Robert Bononno. Minneapolis: University of Minnesota Press, 2010.

Stengers, Isabelle. 'Introductory Notes on an Ecology of Practices'. *Cultural Studies Review* 11, no. 1 (2005): 183–96.

Stengers, Isabelle, Erin Manning, and Brian Massumi. 'History through the Middle: Between Macro and Mesopolitics'. Translated by Brian Massumi. *Inflexions: A Journal for Research-Creation* 3 (2009).

Tambiah, Stanley. *Magic, Science, Religion and the Scope of Rationality*. Cambridge: Cambridge University Press, 1990.

Tan, Min et al. 'Fellatio by Fruit Bats Prolongs Copulation Time'. *PLoS ONE* 4, no. 10 (2009): 1–5.

The Animal Studies Group. *Killing Animals*. Urbana: University of Illinois Press, 2006.

Thompson, John N. *The Geographic Mosaic of Coevolution*. Chicago: University of Chicago Press, 2005.

Thomson, Melanie S. 'Placing the Wild in the City: "Thinking With" Melbourne's Bats'. *Society and Animals* 15, no. 1 (2007): 79–95.

Tidemann, Christopher R. and J. E. Nelson. 'Long-Distance Movements of the Grey-Headed Flying Fox (*Pteropus Poliocephalus*)'. *Journal of Zoology* 263, no. 2 (2004): 141–6.

Tidemann, Christopher R. and Michael J. Vardon. 'Pests, Pestilence, Pollen and Pot Roasts: The Need for Community Based Management of Flying Foxes in Australia'. *Australian Biologist* 10, no. 1 (March 1997): 77–83.

Tidemann, Christopher R., Peggy Eby, Kerryn Parry-Jones, and Michael Vardon. 'Grey-Headed Flying Fox'. In *The Action Plan for Australian Bats*, edited by A. Duncan, B. Baker, and N. Montgomery, 31–35. Canberra: Natural Heritage Trust, 1999.

Tsing, Anna. 'Arts of Inclusion, or, How to Love a Mushroom'. *Australian Humanities Review* 50 (2011): 5–21.

Turner, Lynn. 'When Species Kiss: Some Recent Correspondence between Animots'. *Humanimalia* 2, no. 1 (2010): 60–86.

van Dooren, Thom. 'Care'. Living Lexicon for the Environmental Humanities. *Environmental Humanities* 5, no. 1 (2014): 291–4.

van Dooren, Thom. *Flight Ways: Life and Loss at the Edge of Extinction*. New York: Columbia University Press, 2014.

van Dooren, Thom. 'Invasive Species in Penguin Worlds: An Ethical Taxonomy of Killing for Conservation'. *Conservation and Society* 9, no. 4 (2011): 286–98.

van Dooren, Thom. *The Wake of Crows: Living and Dying in Shared Worlds*. New York: Columbia University Press, 2019.

van Dooren, Thom and Deborah Bird Rose. 'Lively Ethography: Storying Animist Worlds'. *Environmental Humanities* 8, no. 1 (2016): 77–94.

van Dooren, Thom and Deborah Bird Rose. 'Storied-Places in a Multispecies City'. *Humanimalia* 3, no. 2 (Spring 2012): 1–27.

van Dooren, Thom, Eben Kirksey, and Ursula Münster. 'Multispecies Studies: Cultivating Arts of Attentiveness'. *Environmental Humanities* 8, no. 1 (2016): 1–23.

Vardon, M. J. et al. 'Seasonal Habitat Use by Flying-Foxes, *Pteropus Alecto* and *P. Scapulatus* (Megachiroptera), in Monsoonal Australia'. *Journal of Zoology* 253, no. 4 (2001): 523–35.

Vidot, Anna. 'Flying foxes crossing Bass Strait'. *ABC Rural*, 30 August 2010. http://www.abc.net.au/sitearchive/rural/tas/content/2010/08/s2997424.htm

Wadiwel, Dinesh. 'Biopolitics'. In *Critical Terms for Animal Studies*, edited by Lori Gruen, 79–98. Chicago: The University of Chicago Press, 2018.

Walker, Keith F., Jim T. Puckridge, and Stuart J. Blanch. 'Irrigation Development on Cooper Creek, Central Australia – Prospects for a Regulated Economy in a Boom-and-Bust Ecology'. *Aquatic Conservation: Marine and Freshwater Ecosystems* 7, no. 1 (1997): 63–73.

Ward, J. V. and Jack A. Stanford. 'Ecological Connectivity in Alluvial River Ecosystems and its Disruption by Flow Regulation'. *Regulated Rivers: Research and Management* 11, no. 1 (1995): 105–19.

Welbergen, Justin. 'The Grey-headed flying-fox, *Pteropus poliocephalus*'. Accessed 2010. http://www.zoo.cam.ac.uk/zoostaff/BBE/Welbergen/GHFlyingFox.htm

Welbergen, Justin and Peggy Eby. 'Not in my backyard? How to live alongside flying-foxes in urban Australia'. *The Conversation*, 27 May 2016. https://theconversation.com/not-in-my-backyard-how-to-live-alongside-flying-foxes-in-urban-australia-59893

Welbergen, Justin, Carol Booth, and John Martin. 'Killer climate: Tens of thousands of flying foxes dead in a day'. *The Conversation*, 25 February 2014. https://theconversation.com/killer-climate-tens-of-thousands-of-flying-foxes-dead-in-a-day-23227

White, Mary. *Running Down: Water in a Changing Land*. Sydney: Kangaroo Press, 2000.

Wilson, Edward O. *The Future of Life*. New York: Alfred A. Knopf, 2002.

Wohlleben, Peter. *The Hidden Life of Trees*. Translated by Jane Billinghurst. Vancouver: Greystone Books, 2016.

Woodford, James. 'The swingers'. *The Sydney Morning Herald*, 2003.

Wyschogrod, Edith. 'Man-Made Mass Death: Shifting Concepts of Community'. *Journal of the American Academy of Religion* 58, no. 2 (1990): 165–77.

Wyschogrod, Edith. *Spirit in Ashes: Hegel, Heidegger, and Man-Made Mass Death*. New Haven, CT: Yale University Press, 1985.

Yanay, Niza. *The Ideology of Hatred: The Psychic Power of Discourse*. New York: Fordham University Press, 2013.

Anonymous

Bat Conservation & Rescue QLD, Inc. www.bats.org.au

'Bat super immunity to lethal disease could help protect people'. CSIRO, 23 February 2016. https://www.csiro.au/en/News/News-releases/2016/Bat-super-immunity-to-lethal-disease-could-help-protect-people

'Bat wraps #3: Giving injured flying foxes a second chance'. 2011. https://www.youtube.com/watch?v=mDEx0sLgoHk

'Booth v Bosworth ("Flying-fox case 1")'. Environmental Defender's Office (QLD). https://www.edoqld.org.au/flying_fox_1

'Booth v Yardley ("Flying-fox case 4")'. Environmental Defender's Office (QLD). https://www.edoqld.org.au/flying_fox_4

The Centre for Invasive Species Solutions. https://invasives.com.au/

'Cocos palms: A threat to flying-foxes'. NSW Office of Environment and Heritage. http://www.environment.nsw.gov.au/animals/cocos-palm-flying-foxes.htm

'Damage mitigation permits for crop protection'. Queensland

Government: Environment. https://www.ehp.qld.gov.au/wild-
life/livingwith/flyingfoxes/damage-mitigation-permits.html

'Flying Fox Dispersal: Fact Sheet'. Eurobodalla Shire Council. http://
www.riverbendnelligen.com/downloads/flyingfox2.pdf

'Flying fox pups injured in Casino heat wave returned to wild'. *Fraser
Coast Chronicle*, 14 April 2015. https://www.frasercoastchroni
cle.com.au/news/supporters-batty-after-rescue/2605305/

'Flying-foxes'. Eurobodalla Shire Council. http://www.esc.nsw.
gov.au/living-in/about/our-natural-environment/grey-headed-
flying-foxes

'Flying Foxes vs. Freshwater Crocodile | Lands of the Monsoon
| BBC Earth'. BBC Earth, 10 April 2015. https://www.youtube.
com/watch?v=wi30w-Mk2yQ

'NSW govt "destruction of flying fox habitat"'. Echo Netdaily,
26 May 2016. https://www.echo.net.au/2016/05/nsw-govt-ap
proves-destruction-of-flying-fox-habitat/

'Orchard flying fox netting program fully subscribed'. NSW
Government: Department of Primary Industries, 17 March
2016. https://www.raa.nsw.gov.au/__data/assets/pdf_file/0009/
598581/Orchard-Flying-Fox-netting-program-fully-subscribed.
pdf

Pest Tales: Units. http://www.pestales.org.au/units.htm

'QLD councils given power to shift bats'. News.com.au, 1 May
2013. https://www.news.com.au/national/breaking-news/qld-co
uncils-given-power-to-shift-bats/news-story/d0987164c9289e0
4cd8a629e22e4f2ea

'Recovery planning: Restoring life to our threatened species'.
Australian Conservation Foundation, Birdlife Australia,
Environmental Justice Australia, 2015.

'Science and flying-foxes'. Cairns Regional Council. https://www.
cairns.qld.gov.au/community-environment/native-animals/fly
ing-foxes/science-and-flying-foxes

'Tick paralysis & cleft palate'. Tolga Bat Hospital. https://tolgaba
thospital.org/tick-paralysis-cleft-palate/

Tolga Bat Hospital. https://tolgabathospital.org/about-us/

'Tolga scrub'. Tolga Bat Hospital. https://tolgabathospital.org/about-us/tolga-scrub/

Voiceless: The Animal Protection Institute. https://www.voiceless.org.au/

Index

Yanay, Niza, 204
Yarra Bend, 37, 238n61
Yarralin community, in
  multicultural worlds, 70–6, 80,
  82, 86–9, 134
Yilngayarri, Tim, 134, 229–30

Yolngu paintings, 141–2
Yolngu people, 141–4
Young, Riley, 72

zoonotic, 178

EU representative:
Easy Access System Europe
Mustamäe tee 50, 10621 Tallinn, Estonia
Gpsr.requests@easproject.com

www.ingramcontent.com/pod-product-compliance
Lightning Source LLC
Chambersburg PA
CBHW050644270326
41927CB00012B/2865